Topics in Safety, Risk, Reliability and Quality

Volume 35

More information about this series at http://www.springer.com/series/6653

Holly A. H. Handley

The Human Viewpoint for System Architectures

Springer

Holly A. H. Handley
Engineering Management and
Systems Engineering
Old Dominion University
Norfolk, VA, USA

ISSN 1566-0443 ISSN 2215-0285 (electronic)
Topics in Safety, Risk, Reliability and Quality
ISBN 978-3-030-11628-6 ISBN 978-3-030-11629-3 (eBook)
https://doi.org/10.1007/978-3-030-11629-3

Library of Congress Control Number: 2019930368

This Springer imprint is published by the registered company Springer Nature Switzerland AG
The registered company address is: Gewerbestrasse 11, 6330 Cham, Switzerland

To the smartest people I know...
Patrick, Maria and Noella.

Foreword

In 2009, the US Department of Defense (DoD) publicized that Human Systems Integration (HSI) had the potential to enhance overall system effectiveness and lower life cycle ownership costs. While strong HSI acquisition policy and guidance were already in place, the US DoD took the next step in supporting HSI activities by establishing a comprehensive plan to coordinate and manage HSI programs across the services. The DoD concluded that human performance assessments were often not integrated and performed too late to influence the design stages of the system acquisition process; the new policy emphasized early identification of HSI domain-level risks, issues and opportunities for improvement and cost reduction, critical to program success. This included identifying human performance impacts as part of the specification of overall system performance requirements.

HSI had been formally recognized as an element of the systems engineering process in 2004, however there were few methods and tools available to capture the implications of HSI-related design decisions, making it difficult to assess the value of HSI recommendations. Total system performance or total ownership cost in complex technical systems cannot be fully realized and managed without accounting for the human component. Comprehensive and effective integration of the human into the full life cycle systems engineering effort is critical to the design, development, and operation of successful systems. The integration of HSI with systems engineering helps to ensure that the project is executed in a coordinated manner so as to achieve mission success.

When the Joint HSI Working Group (JHSIWG) conducted a needs analysis for HSI, it recommended that the HSI and system engineering communities should undertake a collaborative effort to design system architecture views that would fully support an integrated HSI assessment capability. Human focused views can capture information on human capabilities, constraints, tasks, roles, networks, training, and metrics. They provide a more complete representation of mission effectiveness by including human capabilities and limitations as an integral part of the system design. A Human Viewpoint would enable users to conduct "what if" analyses to support trade-offs across domains. This approach would enable HSI practitioners to most effectively support and integrate with system engineering processes and tools throughout the acquisition life cycle.

The Human Viewpoint was developed by a panel of system engineering and HSI practitioners in 2007. The goal was to develop an integrated set of models, similar to existing architecture viewpoints, that included and organized human data as part of the architecture description. HSI practitioners have long argued that without a viewpoint that focuses on the human component of the system, there is no basis in the architecture for analysis of human issues that may impact multiple aspects of the system. With a viewpoint that captures human considerations, analyses that include the human specifications to adequately operate and maintain the system can be assessed and addressed early in the acquisition process. This ensures efficient and effective use of human resources within the system, ultimately reducing overall system costs. The development of the Human Views has been included in the latest guide for HSI practitioners.

The Human Viewpoint supports HSI's goals of improved integration of humans and systems. Humans play a pivotal role in the performance and operation of most systems, i.e., systems must be supported by sufficient manpower, and personnel must be adequately trained to operate the system. Therefore, the absence of a human perspective in the architecture framework leaves a gap in both the system architecting and acquisition processes. The Human Viewpoint organizes information and provides a comprehensive representation of human capabilities related to expected performance. It provides a basis for decisions by stakeholders by enabling structured linkages from the engineering community to the HSI community. It provides a fully integrated set of products that can be used to inform and influence system design, development, and production processes, facilitates human system tradeoff considerations, and it ensures the human component has visibility as part of the system acquisition process.

This volume, *The Human Viewpoint for Systems Architectures*, provides a comprehensive guide to apply the Human Viewpoint methodology for different types of systems. The implementation of the Human Views supports HSI's goals of optimizing total system performance, reducing life cycle costs, and minimizing risks by ensuring a systematic consideration of the total system throughout the system design process. A successful Human Viewpoint development, completed during the system acquisition process, can result in risk reduction and fewer changes in the mature system. This book can guide both System Engineers and HSI Practitioners to a successful system realization.

Alexandria, VA, USA
Winter 2019

Dr. Beverly Knapp

Contents

Chapter 1
Introduction

Robert Smillie
robert.smillie@cox.net

Abstract This chapter introduces the Human Viewpoint by reviewing the origins of system architecting and architecture frameworks, especially as driven by the US Department of Defense. It recognizes the initial concerns of the Human System Integration community regarding a lack of focus on human limitations and the need for a dedicated viewpoint. It documents the initial workarounds used to augment existing frameworks with human focused data, and assesses the capabilities of alternative frameworks. Finally, it recounts the workshop where the initial Human Views were developed, the precursor to today's Human Viewpoint.

Keywords Human system integration · Architecture frameworks · Human Views

1.1 Background

According to the International Ergonomics Association (2018), "ergonomics is the scientific discipline concerned with the understanding of interactions among humans and other elements of a system, and the profession that applies theory, principles, data and methods to design in order to optimize human well-being and overall system performance." The term "human factors" is generally considered synonymous with the term "ergonomics".

While it can be argued that researchers and analysts have been concerned with ergonomic related problems for over 200 years, the discipline of human factors emerged during World War II as a result of the various design issues and concerns that professionals noted with the operation of manned systems (e.g., airplanes, radar, tanks, etc.). Human factors has always taken a systems approach, however in complex system design and development it was recognized that the role of the human in any system had to be considered from a number of perspectives. The domains of Human Factors Engineering, Manpower, Personnel, Training, Health and Safety, Habitability, and Survivability have to be addressed in an integrated approach, and that approach is known as Human Systems Integration (HSI).

HSI is a systematic process for identifying, tracking and resolving human related issues ensuring a balanced development of both the technological and the human

H. A. H. Handley, *The Human Viewpoint for System Architectures*, Topics in Safety, Risk, Reliability and Quality 35, https://doi.org/10.1007/978-3-030-11629-3_1

aspects of a capability. It is an integral component of systems engineering. Systems engineering is an interdisciplinary field of engineering and engineering management that focuses on how to design and manage complex systems over the system life cycle. In order to ensure system success, HSI has to be a major component of systems engineering.

1.2 Department of Defense Frameworks

In practice, systems engineers use architecture frameworks to describe complex systems. An architecture framework defines a common approach for development, presentation, and integration of system descriptions. The application of the framework contributes to more effectively building interoperable systems and providing a mechanism for understanding and managing complexity. Frameworks capture much more than abstract or functional decomposition of systems. The models capture multiple views of a complex system, which can be integrated to recreate the system. Executable models used to evaluate performance measures can be created from the information captured in these system models.

As a result of information technology and acquisition reform in 1996, the United States Department of Defense Architecture Framework (DoDAF) emerged as the structure for development of a system or enterprise architecture. DoDAF approaches are applicable to large systems with complex integration and interoperability challenges and are used by the engineering and acquisition communities to describe the overall system. Using DoDAF as the basis, similar approaches outside the US evolved, including the Canadian Department of National Defence Architecture Framework (DNDAF) and the United Kingdom Ministry of Defence Architecture Framework (MoDAF).

DoDAF defines different perspectives or viewpoints that logically combine to describe system architectures (DoDAF 2004). DoDAF uses viewpoints to detail a system in terms of its operational capability; it currently defines eight viewpoints that break complex systems into specific categories, including Capabilities, Data and Information, Operational, Project, Services, Standards and Systems Viewpoints. It also includes an All Viewpoint that describes the overall scope of the architecture, as well as support for "Fit for Purpose" views to address specific stakeholder questions. Each of the viewpoints depicts certain architecture attributes. Some attributes bridge two viewpoints and provide integrity, coherence, and consistency to architecture descriptions.

MoDAF was adapted by the United Kingdom Ministry of Defence (MoD) from the DoDAF (MoDAF 2005). The DoDAF Viewpoints were extended into six MoDAF viewpoints. MoDAF added the Strategic View (StV) and the Acquisition View (AcV). The StV consists of viewpoints that articulate high level requirements for enterprise change over time, whereas the AcV consists of views that describe programmatic details to guide the acquisition and fielding processes. The Canadian DNDAF (DNDAF 2010) is also closely based on DoDAF. DNDAF provides a Common View

(CV), Operational View (OV), System View (SV), and Technical View (TV), all similar to the DoDAF viewpoints, but also includes an Information View (IV) and Security View (SecV).

Architecture models are developed in the course of creating a given architecture description and describe the characteristics pertinent to the purpose of the architecture. These models, or views, can take graphical, textual, or tabular form. It is important to distinguish between an architecture viewpoint and an architecture view. A viewpoint represents a perspective on a given architecture, while a view is a specific representation of a particular aspect of that perspective. Thus, a viewpoint will consist of one or more architecture views. At the lowest level of the framework, the architecture data elements are the basic building blocks for inclusion in each architecture model. An integrated architecture insures that data elements defined in one view are the same as the elements in another view. This creates common points of reference, linking together common architecture data elements, ensuring that relationships between the views are maintained.

1.3 Emergence of the Human Views

In the early 2000s, an HSI analyst was assigned to the U.S. Navy's Chief Engineer's office. DoDAF was just beginning to emerge as the mechanism for describing complex systems. What the analyst observed was that the DoDAF did not address viewpoints that captured human functions and activities. It captured the system hardware functions and related activities, but not how the human operator or maintainer was a part of the system. She recognized that without human views, there is no basis in the architecture for analysis of human issues (Adams and Hildebrand 2002). In the 2004 DoDAF Deskbook, the analyst made an initial attempt to represent humans in the then DoDAF products by including the roles and human activities associated with a system. The intent was to promote a set of Human Views as a necessary system architectural sub-view that defines the role of the human in the system and captures the human activities and tasks related to the system. Unfortunately, little was done by the Department of Defense or the individual services to expand that representation and integrate it as a fundamental aspect of the DoDAF. Follow-on analytical efforts in both the United Kingdom and Canada, however, did focus on how to include human activities in an architecture framework.

A detailed assessment of all MoDAF Views was performed in order to identify a list of potential MoDAF shortcomings that would lead to HSI problems if not addressed (Bruseberg and Lintern 2007). The shortcomings noted included: a. HSI trends and standards were not captured; b. human performance metrics, targets, and limitations were not specified; c. human organizational design was insufficiently captured; d. allocation of functions and information requirements specifications could be forced by technology constraints; and e. team activity and team requirements were insufficiently captured. A set of Human Views was suggested that was complementary to the existing MoDAF Views and explicitly specified the HSI elements that need to be considered in the design of socio-technical systems. By identifying specific HSI design elements in relation to the technological elements, HSI analysis, assessment, and management activities can be better related to enterprise design concerns.

The Canadian approach presented an extension to the existing DNDAF in the form of a limited set of human architecture products that specifically targeted decision makers interested in the HSI areas of manpower, career progression, and training (Baker et al. 2006). These domains collectively define how the human component will impact system or capability performance (e.g., mission performance, safety, supportability, and cost). Conversely, the HSI domains also define how the system impacts the human component (e.g., trade structures, skill gaps and training requirements, manning levels, career progression, selection and retention, workload, and morale). Collectively, the proposed HSI supplements were intended to help define and describe the role of the human within a system.

In 2005, the North Atlantic Treaty Organization (NATO) Human Factors and Medicine (HFM) Research Technology Organization (RTO) Task Group (RTG) 155 (NATO RTO HFM-155) was formed (NATO 2010). The task group met over the next five years; one of its objectives was to develop draft characteristics and parameters for "Human Views" to augment the systems architecture products used by systems engineers responsible for the design of complex systems. NATO had followed the lead of the countries developing and defining architecture frameworks and defined its own, the NATO Architecture Framework (NAF). While these frameworks had evolved to include systems engineering concepts, the portrayal of the human as a unique part of the system had not been broached. NATO RTO HFM-155 examined how the human could be better represented within the total system, through the specification of a Human Viewpoint.

The Human Viewpoint explicitly represents the human and documents the unique characteristics humans bring to a system design. It enables an understanding of the human role in system or enterprise architectures. It provides a basis for stakeholder's decisions by linking the engineering community to manpower, personnel, training, and human factors communities. It integrates HSI into the mainstream acquisition and system engineering process by ensuring that human roles are considered early and often. It provides early coordination of task analysis efforts by both systems engineering and HSI teams. A universally accepted Human Viewpoint enables consistency and commonality across service elements, coalition forces, or any large complex system development effort. By capturing the necessary decision data and integrating these with the rest of the architecture framework, the Human Viewpoint provides a more complete set of system data and characteristics.

Using a workshop approach, the NATO panel identified characteristics and parameters for individual Human Views which could be used to augment the systems architecture products required of systems engineers designing a major complex system. Each country had an architecture requirement similar enough in nature and intent to provide the basis for review and analysis, and show gaps in how human roles and requirements were represented.

The purpose of the Human Views is to define the role of the human in the system and to capture the human operator activities, tasks, communications and collaborations required to accomplish mission operations and support operational requirements (NATO 2010). With the Human Views, the role of the human within the system is defined and task activities are described at a level useful for analysis. The necessity of human activities in the system can then be weighed against manpower and training costs associated with human presence. Human characteristics, limitations, and constraints that effect performance can also be considered. The Human Views may be the driver for the systems and technical viewpoints in a human-centered design. Without this view there is no basis in the architecture for analysis of human issues.

1.4 The Human Viewpoint

None of the DoDAF defined viewpoints focus explicitly on human-focused data. By augmenting the system architecture description with a set of Human Views, a more complete set of system data is provided for development and analysis. The Human Viewpoint organizes information into a framework about how the human functions in the system in order to model the impacts of human performance from tasks, personnel, and system resources. It provides a set of models which captures information on human capabilities, constraints, tasks, roles, networks, training, and metrics, which are integrated with a dynamic model used to determine human risk.

HSI support for system architectures is easily facilitated by utilizing appropriate Human Views as well as identifying and providing data for traditional architecture viewpoints. The Human Viewpoint provides a communication medium for HSI and systems engineering to address stakeholder concerns on personnel issues. Knowledge of human skills, capabilities, and limitations, as well as training, safety, and manpower and personnel are applied to system architecture development to ensure human contributions and interfaces to mission performance are appropriately represented.

Use of the Human Viewpoint does not suggest a new type of analysis or a competing process to existing HSI tools and techniques. Defining the Human Views is a process that illustrates the interconnections among system engineering and HSI. System engineering is an interdisciplinary approach and a means to enable the realization of successful systems; systems engineers use architecture frameworks to describe complex systems. HSI is integral component of system engineering, both of which, must be accomplished together to ensure system success.

Architecture frameworks have continued to evolve. While the US continues to rely on its own approach, other countries appear to be moving to a common NATO framework that is information-centric. It divides the framework up into categories of architectural information rather than how the information is presented (NAF 2013). It is not clear that either of these approaches fully integrate the Human Views.

1.5 Summary

The Human Viewpoint provides a methodology to incorporate human capabilities into system development. By treating humans as elements of the system, their knowledge, skills, and abilities can be assessed as attributes of the system, and the interfaces between the technological components and the human components become the integration points. Large, complex systems operate in distributed environments and require the specifications of the technological systems, as well as the social, organizational, task, and skill structures that support the flow of information. During the overall life cycle of systems, the human element is the costliest resource. Systems must be supported by a sufficient number of operators and maintainers who are adequately trained to operate the system in the context of an operational mission.

While the Human View models provide a repository for the required HSI data, the Human Dynamics is necessary to provide the types of analyses required to ensure that the burden on the human component, through workload, training, or other constructs, does not result in poor system design or implementation. Overall, the use of the Human Viewpoint is critical in the architecture framework because it captures all aspects of the human system components.

References

Adams C, Hildebrand G (2002) Human views in the systems architecture framework. Unpublished presentation, Naval Surface Warfare Center, Dahlgren, VA

Baker K, Pogue C, Pagotto J et al (2006) Human views: addressing the human element of capability-based decisions. Presented at the defence science and technology organisation human systems integration symposium, Canberra, Australia

Bruseberg A, Lintern G (2007) Human factors integration for MoDAF: needs and solution approaches. Presented at the 17th annual international symposium, INOCSE, San Diego, CA

DNDAF (2010) Department of National Defence/Canadian Armed Forces architecture framework v1.7. www.img.forces.gc.ca/pub/af-ca/index-eng.asp. Accessed 9 Sept 2018

DoDAF (2004) DoD architecture framework version 1.0, Deskbook. Department of Defense Architecture Framework Working Group, Washington, DC

IEA (2018) What is ergonomics? International Ergonomics Association. https://www.iea.cc/whats/index.html. Accessed 21 June 2018

MoDAF (2005) Ministry of defence architectural framework: technical handbook version 1.0, London, UK

NAF (2013) The NATO architecture framework v3.0. North Atlantic Treaty Organization, Science and Technology, Brussels, Belgium

NATO (2010) Human systems integration for network centric warfare. Technical report, North Atlantic Treaty Organization Research Technology Organization, TR-HFM-155, Brussels, Belgium

Chapter 2
Human System Engineering

Abstract This chapter defines the concept of Human System Engineering and its application to the system engineering process. Including human engineering principals in the system architecting phase of system development ensures the human component is fully considered in the subsequent system design phase. The Human Viewpoint methodology identifies the context of system use, the types of human focused data required, and develops representations to display the data to stakeholders. The methodology is tightly coupled with Human System Integration processes and provides an early assessment of human limitations and constraints.

Keywords Human system engineering · Socio-technical systems · System development

2.1 Introduction

System architecting is performed at the beginning of the system acquisition process in order to explore alternatives, test assumptions and answer stakeholder questions regarding a new system design or system modifications. This book approaches system architecting from a human perspective. It provides a methodology to integrate human concerns into the development and analysis of socio-technical systems. A socio-technical system refers to the human-technology partnership that exists for systems that depend on user interactions. Based on the original NATO Human View developed to augment existing architecture frameworks, the Human Views Framework, or Human Viewpoint, and its accompanying methodology, have evolved as standalone tools to represent human-system concerns for socio-technical systems. This chapter introduces the domain of Human System Engineering and its application of the Human Viewpoint, as well as the relationship between Human System Engineering and Human System Integration.

H. A. H. Handley, *The Human Viewpoint for System Architectures*, Topics in Safety, Risk, Reliability and Quality 35, https://doi.org/10.1007/978-3-030-11629-3_2

2.2 Human System Engineering

Human System Engineering (HSE) is the application of human principles, models and techniques to system design with the goal of optimizing system performance by taking human capabilities and limitations into consideration (DOD 1988). HSE is especially concerned with the human component as a system resource, i.e., what the required skills are, what tasks are assigned, and how roles are defined that can be assigned to specific personnel. A key focus is on the determination of the human role strategy, i.e., the need for different types of personnel based on the system task requirements. The human role strategy is an important design outcome as it determines the implications for manning, training, and ultimately cost (ONR 1998). Thus, the focus of HSE is on often on determining user roles based on sets of tasks and identifying the required knowledge, skills and abilities (KSAs). Including HSE activities in the system architecting process improves the overall system performance by considering human capabilities and limitations early in the design process.

Robotics, automation and artificial intelligence are changing the nature of socio-technical systems and the role of the human. Human interactions are changing from those of an operator to those of a supervisor, and organizational roles are being redefined as tasks are rebalanced between humans and machines. However, as some traditional roles are eliminated, new ones are created, as even in an automated world there will still be a requirement for human workers (PWC 2017). HSE identifies the interactions required between humans, humans with systems, and humans with automation to identify the requirements of the human component in system processes and information flows.

Human focused analyses that occur as part of the HSE evaluations determine the required interactions between users and technology and then evaluate the role assignments based on availability and task loading. The inclusion of the human in the system development is essential to insure efficient processes and data exchange between the technology elements and the human users. These analyses can occur at the individual level (defining individual roles and skills), team or crew level (defining crew composition, work distribution, and human interactions), as well as at the organizational level (overall manpower and personnel requirements).

2.3 Relationship to System Architecting

In the 1990s, the concept of a system architecture was introduced, driven by the rapidly increasing complexity of information centric systems. The purpose of the system architecture was to capture the underlying structure of a system by identifying its functionality, interconnections and core technologies; system complexity could then be managed by identifying and classifying the constituent parts. Architectures were designed to be able to compare the current state of a system ("as-is") to design options for future states ("to-be") in order to develop and manage a migration strategy for the system (DOD 1997). Additionally, the impacts of design decisions on system resources could be evaluated and the trade-offs between system constraints such as risk, cost and schedule could be included in the system acquisition decision making process. The resulting system architecture describes the system through different views that capture its operational, technical, and behavioral aspects. The system architecture description then transitions to the system design, providing the models that can be used to realize the system.

Around the same time as the emergence of system architectures, the field of system engineering matured with the onset of large, complex systems. These systems required both a manager and integrator to ensure that the system was designed and developed to meet the customer's needs while remaining within constraints. System engineering is an interdisciplinary approach and a means to enable the realization of successful systems (INCOSE 2015). The development of the system architecture became the responsibility of system engineers.

Human system engineers are actively involved at the system architecting stage of system development. In this phase, the system concept is developed and stakeholder concerns identified. The human system engineer partners with the system architect to ensure that human considerations are included in the architecture description. Human system analysis done in the system architecting phase ensures that the requirements for a qualified workforce to operate and support the system are in place at system realization. In the system architecting stage, HSE focuses on understanding the capabilities that the system provides, the tasks that need to be performed, the allocation of specific tasks to human roles, the determination of the required knowledge, skills, and abilities, and the constrains imposed by human operator capabilities and limitations.

The goal of the HSE effort is to augment the system architecture description with human-centered models and analyses. These purposeful models inform trade-off analyses between system design, program costs, schedule and overall performance. For example, technology decisions may include increased human reliance on automation, which may impact both human cognitive loading as well as system network management, as networked interactions demand complex information needs. The impact of different system design considerations on manpower, i.e., how many people are needed, personnel, i.e., the types of people needed, and training, i.e. providing specialized skills, can be included in overall system cost projections.

As part of the system architecting team, HSE supports architectural development that integrates human considerations into system design. System engineers and human system engineers are both vested in system success. System engineers facilitate the integration of all sub systems to insure a successful system realization that meets performance as well as cost and schedule objectives. Human system engineers focus on the role definitions, task assignments and personnel requirements, incorporating the human-related specifications into the system description. By participating in the system architecting development, human system engineers ensure human centered principles are incorporated into design decisions. The resulting models and analysis provide a human-centered context to the system, while retaining the same formats and "language" as the system architecting models, providing seamless integration to the overall system architecture description.

2.4 Human Viewpoint Methodology

The Human Viewpoint and its accompanying methodology are a framework and a process to capture socio-technical interactions and human-system requirements. Throughout the system engineering process, inclusion of the human element is essential to ensure understanding of the role the human plays in the performance of the system and to guarantee that appropriate personnel are available to operate the system. The Human Viewpoint provides a framework to capture different sets of human-centered information, referred to as Human Views. They provide a comprehensive representation of human parameters that can be used to inform and influence the system development. The accompanying Human Viewpoint methodology is a tailored approach to collecting the data and performing human system analyses based on a five-step process. The Human Views organize the human-centered information into distinct models, providing a working inventory of human system data. These models can then inform the human-centered design decisions for the system.

The set of Human Views is referred to as the Human Viewpoint; the terms Human Views and Human Viewpoint are used interchangeably. The different views are general categories of data that can be further specified based on the type of system or to address explicit stakeholder concerns; a stakeholder is any entity with an interest in the acquisition, design or deployment of the system. While the framework provides general categories of data for use in providing the system architecture description, the Human Viewpoint is customizable and allows the human system engineer to select the form and format appropriate to the current development for the architectural description. This flexibility allows models to be created for analyzing specific socio-technical issues and providing focused information for stakeholder decision support. The descriptions of the individual Human Views are listed in Table 2.1.

Table 2.1 Individual Human Views (Handley and Smillie 2008)

Human View	Description
Concept	Scope of inquiry for the socio-technical system
Tasks	Human specific activities
Roles	Functional responsibilities defined for the humans interacting with the system
Training	Required knowledge, skills and abilities necessary for the roles to complete tasks
Human network	Human communication patterns to support roles and tasks
Metrics	Human related target values and performance criteria
Constraints	Repository for different classes of human limitations

Table 2.2 Human Viewpoint methodology (Handley and Knapp 2014)

Step	Description
1. Context	Create a data map that identifies the content specific data to be captured
2. Data	Collect context focused data in a repository for the domain of interest
3. Models	Render visual models that depict the data and important relationships
4. Analysis	Use the data and models to evaluate design alternatives or answer queries
5. Fit for Purpose	Provide outcome products that include specific data, models, or analysis results to address stakeholder concerns

The accompanying Human Viewpoint methodology provides the process to describe the stakeholder requirements, determine the level of information granularity required, identify the criterion for implementing the views, provide supporting analyses, and decide the form and format of the rendered models. The Human Viewpoint methodology consists of five steps: Context—Data—Models—Analysis—Fit for Purpose. The Context stage creates a data map based on the area of concern. It identifies pertinent entities and relationships for the different Human View models. The Data stage collects information for each of entities from applicable documentation or other materials; this data usually is formatted as tables that can then be used for a variety of models and analysis. The Model stage creates visual models that can capture the relationships of interests or represent the data in other ways that helps focus on the area of concern. The Analysis stage uses the models, algorithms or other tools to suggest solutions to the stakeholder concerns. Finally the Fit for Purpose stage produces custom products that include the results of analyses to support stakeholder decisions. A summary of the stages of the Human Viewpoint methodology is shown in Table 2.2.

In the system architecting stage, alternative conceptual designs can be proposed for consideration. System engineering decisions often evaluate trade-offs between system costs and other factors. The Human Viewpoint can help evaluate design alternatives based on a personnel perspective. For example, a low-cost design may require personnel from job classifications that are in high demand, potentially offsetting any cost savings. In contrast, a higher cost design may contain self-diagnostics, which will not require any maintenance personnel. Likewise, automation of tasks may reduce the number of operators required as well as requiring lower trained or less qualified operators. By considering these types of trade-offs in the system architecting stage, the design stage can be influenced by the outcomes of the human performance analysis.

2.5 Relationship to Human System Integration

Human Systems Integration (HSI) is the interdisciplinary technical process for integrating human considerations into systems engineering practice (DOA 2015). The goal is to improve overall system performance through human performance analysis throughout the system design process. HSI activities are coordinated with program managers and system engineers to ensure that systems are designed to be compatible with users cognitive, physical, and sensory capabilities. Seven HSI areas of concern, or domains, have been identified as shown in Table 2.3; each of these domains are evaluated as the system design progresses through different system engineering stages. The HSE activities that occur in the system architecting stage, specifically focused on the Manpower, Personnel and Training domains, set up the system for successful HSI activities as it transitions from system architecting to system design and development.

The use of the Human Viewpoint has been integrated with HSI methods; recent guidance for HSI practitioners describes in detail the use of the Human Views as part of a robust HSI system assessment (DOA 2018). HSI practitioners have long argued for a viewpoint that focuses on the human component of a system to support analysis of human issues that may impact multiple aspects of the system. HSI assessments that rely on human focused architecture data include performance analyses that consider the human impact to system performance, cost-benefit analyses that consider the impact of manpower, personnel and training on total costs, and requirement analyses that include the human specifications necessary to adequately operate and maintain the system. Utilizing the Human Viewpoint supports HSI's goals of optimizing total system performance, reducing life cycle

Table 2.3 Human system integration domains (DoD 2017)

Domain	Focus
Human factors engineering	Integrate human characteristics into system definition, design, development and evaluation to optimize performance
Personnel	Determine and select the appropriate cognitive, physical and social capabilities required to operate, maintain and sustain systems
Habitability	Establish and enforce requirements for individual and unit physical environments, personnel services and living conditions to mitigate risks that could impact performance
Manpower	Determine the most efficient and cost-effective mix of military and contract support necessary to operate, maintain and support the system
Training	Develop efficient and cost-effective options that enhance user capabilities and maintain skill proficiencies
Safety and occupational health	Ensure appropriate environmental, safety and occupational health are considered in the design to minimize the risk of illness, disability, injury or death
Force protection and survivability	Assess risks to personnel pertaining to the system design to protect them from threat and accidents

costs, and minimizing risk to personnel by ensuring a systematic consideration of the impact of the materiel design on the human throughout the acquisition process.

Likewise, the HSI standard practice emphasizes the development of the Human Viewpoint when human operators or maintainers interface with the proposed system and capability (SAE 2018). It highlights HSI support for system architecture development, including defining human roles, capabilities, information requirements and operating conditions to ensure human contributions to mission performance are represented appropriately. It recognizes that systems engineers use architecture frameworks to describe complex systems, and by including the Human Views, an architecture framework can serve as a communications medium between HSI and system engineering, both of which must be accomplished together to ensure system success.

2.6 Example

Self-propelled vacuum cleaners are becoming less of a novelty and more of a mainstay in many homes and offices, see Fig. 2.1. These systems operate on a pre-programmed schedule and require minimum human intervention, unless an error occurs or the dust bin is full. In the context of the human operator, the robot changes the functionality of vacuuming from a physical demand task (i.e., pushing an upright vacuum) to a cognitive demand task (i.e., monitoring for alerts). What is the impact of this technology change on the human role and tasks? Initially the system may be assigned to the existing role for the manual vacuuming function, such as the custodian or housing-keeping role. However, since the task demands have changed, a better fit of skills to the technology may be assigning the vacuum function to a front-desk or other administrative role, which can perform the monitoring task. HSE can help understand the changes of roles and responsibilities due to changes in technology and the implications for required skills and reassignment of tasks.

2.7 Summary

Human System Engineering focuses on understanding human roles and tasks within socio-technical systems. As systems change in response to new technologies, the roles and tasking for human operators are being redefined. HSE can be used to determine what personnel can be repurposed to new roles, how to balance tasking among personnel, and how to identify where new skills are needed in order to best use and assign humans resources. In some cases, technology may replace humans, in other cases automation and will become a collaborator. "We will be entering into a new kind of partnership with machines that will build on our mutual strengths, resulting in a new level of human-machine codependence" (IFTF 2011).

The Human Views capture human-centric data and organize the information into a framework in order to model the impacts of human performance from tasks, person-

Fig. 2.1 Human-propelled and self-propelled vacuum cleaners

nel, and system resources. The Human Viewpoint provides a set of models that captures information on human capabilities, constraints, tasks, roles, networks, training, and metrics. These models augment the system architecture description and provide a more complete representation by including human capabilities and limitations as an integral part of the system design.

The Human Viewpoint methodology is part of the system architecting process. Systems architecting is an integrative approach to capture the structure and behavior of a system, with the goal of managing complexity by creating models of the proposed system for communication among stakeholders. The Human Views development occurs during the system architecting process with the goal of capturing the human component, and its relationship to the developing system, for inclusion in the architecture description for stakeholder discussions and decisions.

References

DOA (2015) Soldier-materiel systems human systems integration in the system acquisition process. Department of the Army Regulation 602–2, Washington, DC
DOA (2018) Guide for human systems integration (HSI) in the system acquisition process. Headquarters Department of the Army, Pamphlet 602-X, Washington, DC
DOD (1988) Manpower, personnel, training, and safety (MPTS) in the defense system acquisition process. DoD Directive 5000.53, Washington, DC
DOD (1997) C4ISR architecture framework, version 2. C4ISR Architecture Working Group, Department of Defense, Washington, DC
DOD (2017) Operation of the defense acquisition system. DoD Instruction 5000.02, Change 3, Washington, DC
Handley H, Knapp B (2014) Where are the people? The human view approach for architecting and acquisition. Def Acquis Res J 21(4)(71):851–874
Handley H, Smillie R (2008) Architecture framework human view: the NATO approach. Syst Eng 11(2):156–164
IFTF (2011) Future work skills 2020. Institute for the Future for Apollo Research Institute, Palo Alto, CA
INCOSE (2015) Systems engineering handbook, 4th edn. The International Council on Systems Engineering, San Diego, CA
ONR (1998) Human engineering process. Technical Report, SC-21 S&T Manning Affordability Initiative, Office of Naval Research, Washington, DC
PWC (2017) Workforce of the future: the competing forces shaping 2030. Price Waterhouse Cooper, www.pwc.com/people. Accessed 15 Jan 2018
SAE (2018) Standard practice for human systems integration. SAE International. www.sae.org/standards/content/sae6906/. Accessed 15 May 2018

Chapter 3
Architecture Concepts

Abstract This chapter provides the foundations of system architecting. It describes in detail the original three viewpoints of the Department of Defense Architecting Framework and the focus on government systems. It includes descriptions of the updates to the framework and the current set of eight viewpoints, as well as the custom, Fit for Purpose views. Finally the chapter identifies the concept of capability based system acquisition, and the close ties to the architecting process. This process can be expanded to include the requirements of socio-technical system and the data captured in the Human Views.

Keywords System architecting · Architecture viewpoints · Capability-based acquisition

3.1 Introduction

Systems architecting is a method to fully describe the configuration of an envisioned (or existing) system. Systems architecting is a descriptive approach, with the goal of managing complexity by creating models of the system for communication among stakeholders. Creating an architecture occurs at the beginning of the system development process; decisions made during the architecting phase provide the basis for the design and detailed technical planning that occurs in subsequent system engineering phases. The outcome of the system architecting stage is an architecture description; it can be used to evaluate system alternatives, capture the overall system concept, and set the baseline for the design work to follow.

3.2 System Architectures

A system represents a collection of components organized to accomplish a specific function, i.e., a set of interrelated components that must work together to achieve some common purpose (Blanchard and Fabrycky 2010). System engineering

H. A. H. Handley, *The Human Viewpoint for System Architectures*, Topics in Safety, Risk, Reliability and Quality 35, https://doi.org/10.1007/978-3-030-11629-3_3

is required for large systems in order to manage and integrate the diverse subsystems into a functional system. System engineers provide detailed system requirements that drive the design of system elements and then manage the development of the system over its lifecycle.

System architecting is most prevalent in government and other large system developments that have a formal acquisition process to procure systems that provide new capabilities. System architecting is the initial stage of the system engineering process and focuses on fully developing the system concept by defining the system scope, capturing user requirements, identifying potential areas of concern, and evaluating alternatives before committing to a specific design. The system architect develops a suitable system baseline that meets the stakeholder requirements which provides the basis for the system engineer's detailed technical design. The resulting architecture description provides the configuration of the system; an architecture is defined as "the fundamental organization of a system embodied in its components, their relationships to each other, and to the environment, and the principles guiding its design and evolution" (IEEE Standard 1471-2000).

The system architecture process focuses on iteratively formalizing the system concept in order to provide a description of a complex system. Initial architecture descriptions may be created at a rather abstract level and then revised to a greater level of detail as the specifics of the system are decided. Data about the developing system is identified and used in different models or visualizations to communicate information about the emerging system in the context of the stakeholder requirements. Missing information is continually identified and resolved, and opportunities for alternative solutions are evaluated. Architecture development has been noted to be both an art and a science, as it requires heuristics and intuition, as well as technical expertise, to translate the stakeholder vision into a detailed system concept (Maier and Rechtin 2000).

System architecting provides the information about the system in the form of organized sets of data that can be used to present the system from different perspectives. When a template for system data is created, it is referred to as a "model". When the model is populated with data from a specific system, it is referred to as a "view". Sets of views from a common perspective are called a "viewpoint." The collection of all the views and viewpoints created for a system is defined to be the architecture description. This partitioning of the data helps manage the system complexity by collating similar and related system information, which facilitates providing decision makers with data focused on specific areas of concern.

The original architecture descriptions developed in the late 1990s focused on developing three viewpoints: The Operational, System, and Technical viewpoints, as shown in Fig. 3.1. The Operational views focused on describing the user tasks and information processes, i.e. the system operations; the Systems views focused on the sub system functionality and interfaces, i.e., the components, and the Technology views described the hardware implementation requirements, i.e., the design standards (DoDAF 2004). While each of the viewpoints described an independent aspect of the system, i.e. operational activities, system functions, and technology standards, the resulting architecture description was considered "integrated", as common data

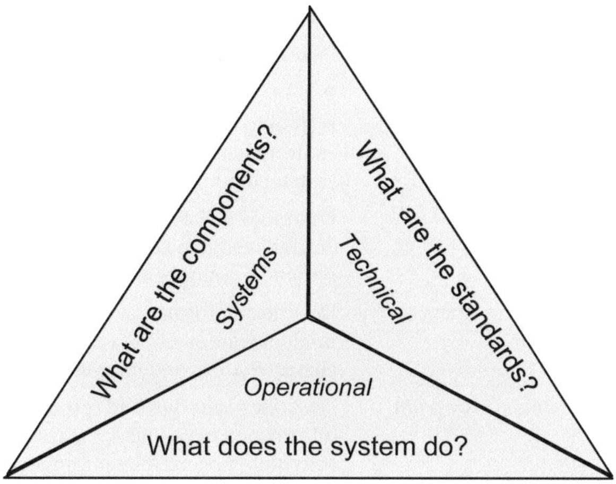

Fig. 3.1 The original three viewpoints architecture approach

appears in each view to ensure consistency across the architecture description. The triangle depiction in Fig. 3.1 represents this integration of data between the three viewpoints.

The resulting architecture description captures the definition of the system concept. It provides a baseline for the detailed development by describing the system elements and relationships from different perspectives. This allows for discussion and analysis to generate a final representation of the system. An architecture is fundamentally about creating a coherent model that enables effective decision-making in order to generate the requirements for a system that can be designed, implemented, and deployed.

3.3 Architecture Frameworks

An architecture framework enables system architecting by providing the taxonomy with which to develop model templates, collect system data and render the views. A framework is a standard that defines all the potential elements and relationships in the architecture description and a set of viewpoints that depict these elements from different perspectives. Different methodologies and tools have been created to assist system architects through the process of collecting system data and developing different views. The methodology is separate from the architecture framework itself, as different methods and tools can support the same framework, and architecting processes can vary across system architecting efforts.

There are many different architecture frameworks available, including the Zachman Framework, the Federal Enterprise Architecture Framework (FEAF), and The Open Group Architecture Framework (TOGAF), as a few examples (NDIA 2013). However, the original Human Viewpoint was derived from the U.S. Department

Table 3.1 DoDAF viewpoint descriptions (DoDAF 2010)

Viewpoint	Description
All	Provides information about the entire architectural description, including the subject area, the timeframe, the operational environment, as well as terminology definitions for use across the architectural description
Capability	Describes a vision for performing specified activities to achieve desired resource states under specified standards and conditions, providing a strategic rationale for the described architecture
Data and information viewpoint	Describes information needs, data requirements, and the implementation of data elements, such as the attributes, characteristics, and inter-relationships of exchanged data
Operational viewpoint	Describes activities and resources, including the types of information exchanged, the frequency of such exchanges, and the activities supported by information exchanges
Project viewpoint	Describes how programs are grouped in as a coherent portfolio of acquisition programs and provides a way of describing the organizational relationships between multiple acquisition programs responsible for delivering different capabilities
Services viewpoint	Describes services that provide or support operational activities and traces service activities and resources to the requirements
Standards viewpoint	Describes the minimal set of rules governing the arrangement, interaction, and interdependence of systems and system parts by identifying the technical systems implementation guidelines
Systems viewpoint	Describes system activities and resources that support operational activities

of Defense Architecture Framework (DoDAF). DoDAF provides a standard set of conventions for capturing architectural data. It establishes data element definitions, rules, and relationships through a meta-model. It provides a set of predefined model templates with which to render sets of architecture views. The views are logically organized into collections to provide viewpoints. Each viewpoint has a particular system focus, and the corresponding sets of models are defined similarly for each viewpoint, i.e., high level summary information, narrowly focused information for specific purposes, and information about how aspects of the system are interconnected.

The original DoDAF definition consisted of three main viewpoints: The Operational, System and Technical viewpoints. Over several revisions, the current version of DoDAF consists of eight viewpoints, these are listed in Table 3.1. Note that while the architectural description is the composite of the completed viewpoints, it is not necessary to complete all views for each viewpoint, or all viewpoints for the architect description. Depending on the system and the stakeholder concerns, the system architect can customize the architecture description to a subset of the full DoDAF specification.

Using an architecture framework enables greater reuse of architectures and the architectural information, as well as allowing for comparison among architectures based on a common representation. The key benefits of using the DoDAF framework include the definition and standardization of key terms, the use of viewpoints that are integrated across common data, and the focus on expressing the architecture data in ways that supports many stakeholder interests. DoDAF is an integrated architecture, as many views have a mapping between elements across different viewpoints that supports consistency throughout the architecture. This integration provides common points of reference linking together different architecture views through common data elements. Additionally, the use of DoDAF supports Fit for Purpose use, which allows views or models to be created outside the prescribed set. Fit for Purpose models allows the data to be presented specific to stakeholder requirements and emphasizes using architectural data to support analysis, i.e., the generation of graphical representations that can be used to support system specific decision-making.

3.4 System Acquisition

System acquisition is an engineering management process within systems engineering in the context of a regulation framework. An acquisition process is used to manage the procurement of technologies that support specified capabilities for governments or other large organizations. A capability is defined as the ability to execute a specified course of action (CJCS 2012). Capability-based acquisition shifts the focus of acquisition away from attempting to completely describe a system's performance requirements early in the concept definition stage. Instead, capability-based acquisition describes a shortfall in current capabilities, and then devotes funds to developing the technologies that may resolve this shortfall through the system architecting process.

The Joint Capabilities Integration and Development System (JCIDS) is the U.S. Department of Defense (DoD) regulation framework which defines the acquisition requirements and evaluation criteria for potential DoD systems. JCIDS is intended to guide the development of requirements for future acquisition systems to reflect the needs of military services by focusing the requirements generation process on needed capabilities. A JCIDS analysis is completed in the pre-acquisition stage and is composed of a structured, four-step methodology that defines capability gaps, capability needs, and approaches to provide these capabilities within a specified functional or operational area (CJCS 2012); these steps are shown in Table 3.2.

The definition of the JCIDS analysis can also be expanded for socio-technical systems to include the definition of human-focused data and concerns—these are shown in the right column of Table 3.2 (Baker et al. 2006). The JCIDS process presents an opportunity to address manpower, personnel and training, and other socio-technical concerns required by the conceptual system. The Human Viewpoint applied in the system architecting stage for capability-based acquisition can be used to define the socio-technical data required support the JCIDS process.

Table 3.2 JCIDS four step process (DoD 2013)

Step	Description	Output	Socio-technical requirements
Functional area analysis (FAA)	Characterize and prioritize the capabilities, operational tasks, and conditions required to accomplish military objectives	Tasks to be accomplished to achieve objectives	Include the roles for operators, maintainers, and support personnel, and identify critical tasks that will be assigned to humans
Functional needs analysis (FNA)	Assess the ability of current capabilities to accomplish objectives in order to define new capabilities that are aligned with strategic priorities for which solutions must developed	List of capability gaps	Include the assessment of the ability of current personnel to accomplish the identified tasks and determine personnel inventory gaps
Functional solution analysis (FSA)	Identify candidate solutions for filling capability gaps, including non-materiel changes, changes in quantity of existing materiel, product improvements to existing materiel or facilities	Potential integrated approaches to capability gaps	Include the identification of changes in manpower, personnel and/or training, and minor human factors engineering changes that could be made to meet all or part of the capability gap
Post independent analysis (PIA)	An independent analysis of approaches to determine the best fit for potential solutions	Initial capabilities document	Include personnel related issues in potential solutions

3.5 Example

For legacy systems, DoDAF architecture descriptions are often created to capture the "as-is" system in order to provide a baseline for comparison with "to-be" models of the system with the desired improvements. For example, the Commander's Daily Update Brief is a status brief that provides the readiness and operational assets throughout a command, with a focus on the previous 24 h and the next 24 h (Handley and Heacox 2005). The development process that produces the brief includes analyzing data sources, creating Microsoft Power Point slides, and numerous review cycles. The high-level functions and personnel interactions are shown in Fig. 3.2.

The current implementation, or as-is system shown in Fig. 3.2, produces the brief through a manual, staff intensive process that results in static information which is

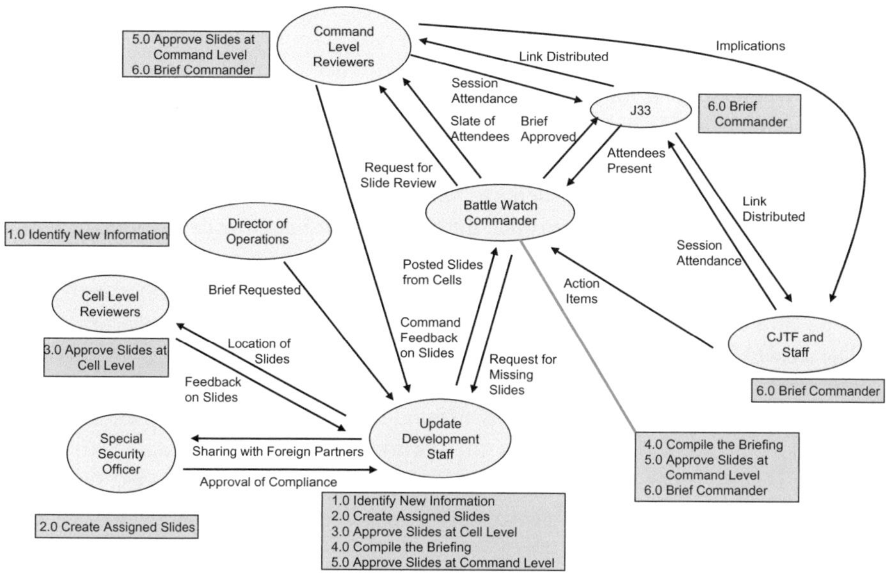

Fig. 3.2 Commander's daily briefing development process

often several hours old. Personnel have to visit Web sites, review text messages, and access databases to retrieve data; the staff then manually transcribe all of the obtained data into a PowerPoint briefing. This labor-intensive brief development process consumes staff members working the night shift, while the day shift's personnel devote the morning (approximately 0600–1100) to its production: editing, multiple briefing evolutions, dry runs and presentation (Pester-DeWan et al. 2003). Additionally, staff working on multiple shifts may result in a brief with data inconsistencies.

A capability-based assessment identified the need for a Web-enabled solution that would integrate and automate the processes required to assemble the brief. The Integrated Interactive Data Briefing Tool (IIDBT) is a system that can automate the data gathering process using Web services that pull data directly from authoritative sources. Additional compatible applications extract selected information from source data and paste it into formatted PowerPoint templates. The commander's staff can continue to update the brief before it is dynamically converted for the final presentation. Including the IIDBT technology with the Commander's Daily Update Brief process represents the desired, or to-be system.

The models produced by the system architecting process for the Commander's Daily Update Brief production cycle for the as-is system can be used evaluate the efficiency of the baseline system. This provides a foundation for determining the projected time and staff savings when integrating new technologies into the brief production cycle. Based on the analysis, the IIDBT is predicted to save the staff an estimated 3.5 h a day while at the same time allowing them to present more current information (Higgins and Hall 2004).

3.6 Summary

Systems architecting is an iterative approach to specify the structure and behavior of a planned system. A system architecture identifies the different components and the relationships between them that contribute to the overall design of the system. The system architecting process helps to manage complexity by using an architecture framework to structure the system data; the resulting architecture description provides a summary of the current or proposed system. Architecture frameworks, including the Human Viewpoint, produce conceptual models that communicate with stakeholders by rendering the data in models or views that support decision making.

The DoDAF architecture viewpoint provides a standard for the collection and representation of system data and offers a consistent way to depict architecture information. It provides predetermined sets of models and viewpoints to present data to stakeholders while also allowing Fit for Purpose presentations to address specific concerns. DoDAF use is mandated in the DoD acquisition process; its use facilitates the acquisition of new capabilities by identifying the relevant components, interactions, and the parameter values necessary to characterize the system baseline. While it is clearly aimed at military systems, DoDAF has broad applicability across the private and public sectors.

References

Baker K, Pogue C, Pagotto J et al (2006) Human views: addressing the human element of capability-based decisions. Presented at the defence science and technology organisation human systems integration symposium, Canberra, Australia

Blanchard B, Fabrycky W (2010) Systems engineering and analysis, 5th edn. Prentice Hall, Upper Saddle River, NJ

CJCS (2012) Manual for the operation of the joint capabilities integration and development system. Chairman of the Joint Chiefs of Staff, Washington, DC

DOD (2013) Operation of the defense acquisition system. DoD instruction 5000.02. Department of Defense, Washington, DC

DoDAF (2004) DoD architecture framework version 1.0, Deskbook. Department of Defense Architecture Framework Working Group, Washington, DC

DoDAF (2010) The DoDAF architecture framework version 2.02. U.S. DoD Deputy Chief Information Officer. http://dodcio.defense.gov/Portals/0/Documents/DODAF/DoDAF_v2–02_web.pdf. Accessed 1 Nov 2018

Handley H, Heacox N (2005) Department of Defense architectural framework products for the commander's daily update brief process, baseline version. Pacific Science & Engineering Final Report, San Diego

Higgins E, Hall J (2004) The IIDBT meets the demands of modern warfare with speed and accuracy. Chips Magazine July–September 2004

IEEE (2000) Recommended practice for architectural description of software-intensive systems. IEEE Standard 1471-2000, September 2000

Maier M, Rechtin E (2000) The art of systems architecting. CRC Press, Boca Raton, FL

NDIA (2013) Use of architectures for integration success. National Defense Industry Association, System Architecture Committee, Task 4 White Paper

Pester-DeWan J, Moore R, Morrison J (2003) Knowledge engineering for command and control transformation. Technical Report, Pacific Science & Engineering, San Diego, CA

Chapter 4
A Socio-technical Architecture

Abstract This chapter introduces the range of socio-technical systems and their interest for system engineers. It reiterates the need to include human focused data at the system architecting stage, in contrast to the traditional practice of leaving the human-system design for the final stages of system development. It notes the benefits of augmenting frameworks with the Human Viewpoint and the opportunity to improve system performance by capturing the contributions of the human component. The chapter details the Human Views as they were originally designed as well as the modifications to the views over time. The Human Viewpoint is an integrated viewpoint, with associations to both the Operational and System Viewpoints.

Keywords Socio-technical analysis · Human Viewpoint · Integrated architecture

4.1 Introduction

The term socio-technical system is used to describe a system that has both a human and a technological component. This implies that the system requirements include aspects of the user interfacing with the system or parameters describing operator decision nodes embedded within the system. In most system performance evaluations, the actions of the system's user or the execution of operator functions impact the measures used to evaluate the system. Traditional system architecting efforts focus on capturing the technical system within a framework, however these frameworks do not include views that identify the human variables that influence the system design and resulting performance. The Human Viewpoint addresses the need for an architecture framework that includes the concerns of a socio-technical system. The Human View models can be used to collect and organize social and technical parameters in order to understand the way that humans interact with other elements of the system. The socio-technical analysis can help understand how the people, technology, and work process come together as a comprehensive system and identify the social and technical limitations.

H. A. H. Handley, *The Human Viewpoint for System Architectures*, Topics in Safety, Risk, Reliability and Quality 35, https://doi.org/10.1007/978-3-030-11629-3_4

4.2 Socio-technical Systems

A socio-technical system refers to the human-technology partnership that exists for systems that depend on user or operator interactions. These systems may have a variety of types and levels of human interfaces, as shown in Fig. 4.1. The systems include wearable systems that sense user inputs, remote controlled systems, that may not be co-located with the user, systems that contain the human, such as self-driving cars, systems composed of teams of humans, such as a space shuttle launch, as well as more traditional types of human interface systems.

Socio-technical systems are often associated with the interaction of operators and technology through work processes (Cummings and Worley 1997). Note, when humans are considered an integral function of the system's functionality, the term "operator" is used; when the human is external to the system, the term "user" is used. Socio-technical systems require an architecture framework where the human focused models are "nested" within the greater system architecture framework in order to perform the socio-technical analyses. These analyses help understand how the people, technology, and work process come together as a comprehensive system, as well as to identify social and technical limitations. Figure 4.2 represents the work process, "Create Assigned Slides", which is one of several sub processes of the Commander's Daily Update Brief described in the example of the previous chapter. This sub process consists of analyzing data sources, identifying pertinent information, and creating Microsoft Power Point slides to communicate information. In Fig. 4.2, the hexagon shapes represent technology assisted tasks and the square shapes represent operator decision nodes.

Traditionally, the system architecting process focused on the functionality (what the system should do) and the corresponding technology (how the system should do it) aspects of the system. However, as the use of architecture descriptions matured, it became apparent that capturing only the system data was not enough to accurately describe both the operation and performance of the system. It had been common practice to leave the engineering of the human component to later in the system

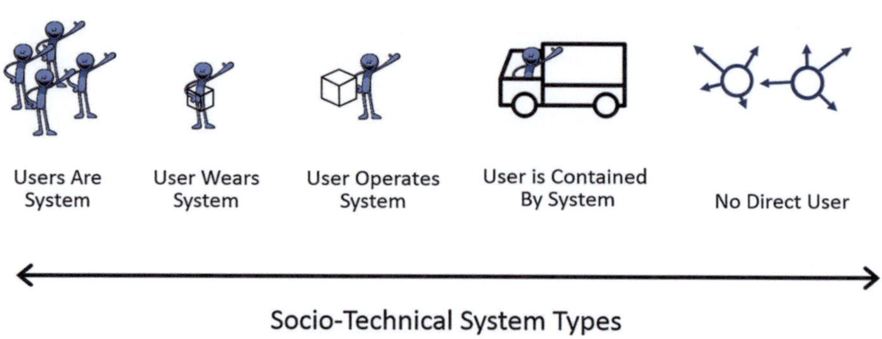

Fig. 4.1 Spectrum of socio-technical systems

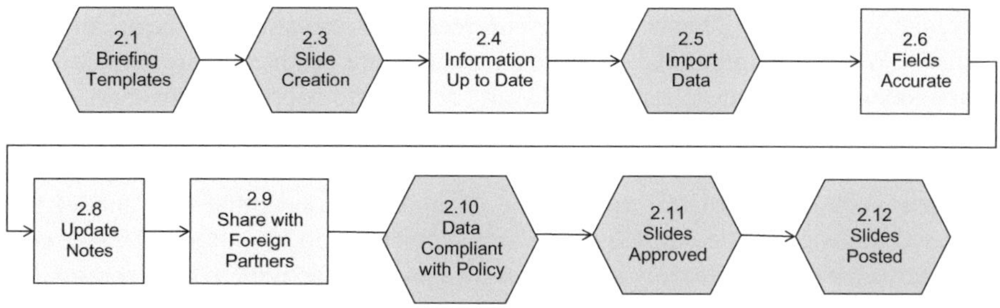

Fig. 4.2 The commander's daily update brief socio-technical work process

design process; during the functional decomposition stage of system architecting, certain functions were noted as human tasks and deferred for later development. However, in order to improve overall system performance and to reduce overall design costs, the architecting process and framework should include human capabilities and constraints in the system architecture description. Early accounting of human abilities, constraints and limitations can prevent future changes due to misperceptions of the human requirements for the system and improve HSI domain assessments during system realization, preventing late stage design changes and work-arounds.

4.3 DoDAF and Human Focused Data

The original Department of Defense Architecture Framework (DoDAF) included three main viewpoints: Operational, System and Technical. There was no viewpoint regarding the human aspects of the system. However, there was acknowledgement of the need to represent the role of humans in the architecture, and some instructions were created to include "Human-Centered Supplementary Architecture Information" in four existing DoDAF products (DODAF 2004). For example, in the Operational Viewpoint, the Organizational Relationships Chart could include human role definitions as part of the organizational structure, and the Operational Activity Model could be used to define the human functions in relation to the operation of the system. From the System Viewpoint, the Systems Functionality Description could include the breakdown of functions performed by humans and those performed by systems, and the Operational Activity to Systems Function Traceability Matrix could include the human activities in the mapping to operational activities. However, this fragmented approach did not provide a cohesive viewpoint for examination and analysis of the impact of the human component on the system design.

The release of DoDAF version 2.0 presented a new philosophy on system architecture descriptions, emphasizing a focus on gathering architecture data pertinent to decision makers' needs rather than an all-encompassing effort to produce the complete range of architecture views (DoDAF 2010). It also encouraged the use of architecting tools to maintain consistency of data and to pursue a model-based approach to system architecting. This was further enforced by the renaming of the architecture artifacts: What were previously known as architecture products were now referred to as models and could be rendered in various ways to support understanding and reasoning about the architecture.

However, DoDAF version 2.0 still did not include a dedicated Human Viewpoint. The new version of DoDAF did include an updated conceptual data model that was used to describe the entities of the architecture description. One of these entities, "performer", refers to any entity that completes an activity, which can include humans as well as technology (DoDAF 2010). With this updated definition, some of the Operational Viewpoint models can be adapted to include a more human centric perspective. For example, the Operational Node Connectivity Description can include nodes identified as human roles performing the corresponding activity in the Operational Activity model. While this method may be sufficient to address certain areas of concern, it would be difficult to assemble the complete collection of human focused data required to perform analyses for socio-technical system evaluations.

A separate, dedicated Human Viewpoint provides the same advantages that the other existing architecting viewpoints provide—allowing the architect or decision maker to focus on a particular area of interest within the context of the overall system. The approach of integrating the human into the Operational Viewpoint as purported by DoDAF version 2.0 does not allow a complete and separate representation of the impact of the human component of the system. A human viewpoint facilitates the collection of human specific data that can be used to describe the capabilities and limitations of the human as part of the architecting process, and then integrate human considerations and analysis into the system architecture description, essential for consideration in the overall system design decision process.

4.4 The Human Viewpoint Development

The Human Viewpoint was developed by a panel of system engineers and HSI practitioners (NATO 2010). Originally termed the "NATO Human View", it is now referred to simply as the Human Viewpoint, in line with the DoDAF version 2.0 terminology.[1] The panel's goal was to develop an integrated set of models, similar to the existing architecture viewpoints that included and organized human data as part of the architecture description (Handley and Smillie 2008). The Human Viewpoint documents the unique implications humans bring to the system design and enables consistency and commonality in representation, consistent with the other viewpoints. With a viewpoint that captures human focused data, human considerations can be evaluated early in the acquisition process, along with its technical counterparts.

The development of the Human Viewpoint provides the ability to capture the human operators as unique components in the system architecture development,

[1]There is also a "Human View for MoDAF" (the United Kingdom's Ministry of Defence Architecture Framework) which is similar to the Human Viewpoint, but not identical.

instead of as specialized system types as proposed in DoDAF version 2.0. This provides a more representative view of the role the human in the system, ensuring efficient and effective use of human resources, and allows for improved human-system trade off considerations. The human data, collected within the framework of the corresponding system data, provides linkages to other viewpoints facilitating a robust socio-technical system analysis.

A dedicated Human Viewpoint also supports HSI's goals of improved integration of humans and systems. Humans play a pivotal role in the performance and operation of most systems, i.e. systems must be supported by sufficient manpower and personnel must be adequately trained to operate the system. The Human Viewpoint provides insights to system designers on human roles and responsibilities and it provides a mechanism to explicitly highlight organizational role changes. This can serve to influence the architecture from a people perspective and identify the impacts of the system on the existing workforce, as well as assist in the planning of needed workforce development.

4.5 The Human Views

The original development of the Human Viewpoint focused on designing a set of products that would capture specific human system data to augment the rest of the architecture description, but would also correspond to the products of the other viewpoints described in DoDAF version 1.0. This would provide consistency across viewpoints and provide for easier integration across the views. The initial design of the Human View identified a series of models that focused on human roles and activities, their technology-based interactions, as well as enablers and constraints due to manpower, training, and human factors issues. Additionally, an eighth product, Human Dynamics, was defined: Its content was recommended to be a simulation model or other analysis of the impact of the human on system performance which could be used to compare alternative human-system configurations. The set of eight products that were defined for the original NATO Human View are shown in Table 4.1 (Handley and Smillie 2008).

Additionally, the Constraints (HV-B), was further decomposed into sub-products. The sub-products of the HV-B are shown in Table 4.2 (Handley and Smillie 2008).

At that time, all of the views within the DoDAF were referred to as products; with the advent of DoDAF version 2.0 products became known as models, and when populated with system specific data, became views of the system. Additionally, all of the views had an identifier based on a two-letter abbreviation of the viewpoint, followed by an index number starting at "1". The Human Viewpoint products were given the same two letter viewpoint abbreviation (HV), but were given an index starting with "A". While they were designed to be similar to the existing products, there was not a one-to-one correspondence with the layout of the products with any of the existing views. In order to avoid confusion, they were given an index of A—H

Table 4.1 Original set of Human Viewpoint products (Handley and Smillie 2008)

Identifier	Name	Description
HV-A	Concept	A conceptual, high-level representation of the human component in the enterprise architecture
HV-B	Constraints	Sets of characteristics that are used to adjust the expected roles and tasks based on the capabilities and limitations of the human in the system
HV-C	Tasks	Descriptions the human-specific activities in the system
HV-D	Roles	Descriptions of the roles that have been defined for the humans interacting with other elements of the system
HV-E	Human network	The human-to-human communication patterns that: occur as a result of ad hoc or deliberate team formation, especially teams distributed across space and time
HV-F	Training	A detailed accounting of how training requirements, strategy, and implementation will impact the human
HV-G	Metrics	A repository for human-related values, priorities and performance criteria, and maps human factors metrics to any other Human View elements
HV-H	Human dynamics	Dynamic aspects of human system components defined in other views

Table 4.2 Original set of constraints sub products (Handley and Smillie 2008)

Identifier	Name	Description
HV-B1	Manpower projections	Illustrates predicted manpower requirements for supporting present and future projects that contribute to larger capabilities. Provides manpower forecasting to allow initial adjustments in training, recruiting, professional development, assignment and personnel management
HV-B2	Career progression	Illustrates career progression as well as the essential tasks, skills, and knowledge (and proficiency level) required for a given job. Addresses impacts of alternative system and capability designs on career progression
HV-B3	Establishment inventory	Defines current number of personnel by rank and job within each establishment. Supports forecasting of trained effective strength and predicting number of people that must be trained, recruited, etc., to fill gaps required for out years
HV-B4	Personnel policy	Defines the various department policies dealing with (governing) HR issues. Ensures that personnel are fairly considered, properly treated, well looked after and supported in a legal, moral and ethical manner while employed

(continued)

Table 4.2 (continued)

Identifier	Name	Description
HV-B5	Health hazards	Considers the design features and operating characteristics of a system that can create significant risks of illness, injury or death. Aims to eliminate, minimize or control both short- and long-term hazards to health that occur as a result of system operation, maintenance and support
HV-B6	Human characteristics	Considers the physical characteristics of an operator, and movement capabilities and limitations of that operator under various conditions. Aims to compare operator capabilities and limitations with system operating requirements under various conditions to improve system capabilities

instead. Overtime, these identifiers have been dropped and the views are identified simply by their name.

The original Human Viewpoint designed in 2007 was the result of a collaborative effort and tried to balance the inputs and views of all of the panel members. Additionally, the product definitions were purposely kept very vague; it was desired that each view could be configured and populated in multiple ways depending on the data available and the analysis being performed. The viewpoint has matured over the last decade as it has been implemented for different systems. As a result, some adjustments and revisions to the original views have been made that have resulted in a more concise viewpoint and improved its usability.

Both specific modifications to the Human Views, as well as subtle definition changes to the original broad categories of human focused data are listed in Table 4.3. Note that the Human Dynamics has been dropped as a separate view—the data from the analyses performed as part of the architecture development is included in the models as per a Fit for Purpose development.

Until recently, the Human Views were completed on an as needed basis, separate from the main architecture development. However, the Human Viewpoint can be considered and implemented as Fit for Purpose views. "Fit for Purpose" is used to describe a set of views that are purposely focused to respond to stakeholder concerns. These customizable views are created to address specific decision maker questions and often provide results of requested analyes. The Human Views can be used to address questions regarding the constraints and limitations of the socio-technical system.

Table 4.3 The present Human Views

Name	Description
Concept	The context and scope for the human system based on high level scenarios or use-cases
Constraints	Limitations that impact the ability of personnel to assume different roles or complete system tasks
Tasks	Descriptions of the human-specific activities performed within the context of interest
Roles	The duties and responsibilities that are assigned to specific personnel
Human network	The interactions and information exchanges required to support and complete task processes
Training	Assessment of required knowledge, skills and abilities or other qualifications required for a role or task
Metrics	Performance standards used to evaluate the ability of personnel to adequately perform assigned tasks

4.6 An Integrated Viewpoint

The Human Viewpoint is an integrated viewpoint. This means that there are interrelationships between both the individual Human Views and between the Human Views and views from other architectural viewpoints. While the Human Views have sometimes been described as "buckets" of individual sets of human focused data, there is a correspondence between the different types of data. Figure 4.3 shows the original relationship diagram for the Human Views. The Concept focuses the Human View development on specific Roles that have been identified for the system. The Roles are assigned to Tasks and interact in the Human Network to exchange information and complete task processes. The Roles may receive Training on system qualifications and certain Training is required for the assignment to tasks. Metrics provide target values for Tasks, and are used in a Dynamics model to evaluate performance. Additionally, Constraints provide both personnel and task limitations that impact task performance. While this presents the general flow through the viewpoint and suggests relationships between the data, it is important to note that it does not limit the relationships between the views to those indicated in the figure.

While the development of the Human Viewpoint often results in a set of "self-contained" Fit for Purpose models, the real value of the Human Viewpoint is as an integrated viewpoint within an architecture framework. The original design of the Human Views purposely considered the relationship of the individual views to the views of other existing viewpoints. Figure 4.4 shows the Human Viewpoint as it was designed to be integrated with DoDAF Operational and System Views. The Concept provides a human focused refinement of the Operational Concept addressing specific stakeholder concerns. The Operational Concept describes the mission of the architecture, often with regard to the interaction of the architecture with its environment. The Tasks decompose the high-level Operational Activities to tasks that can

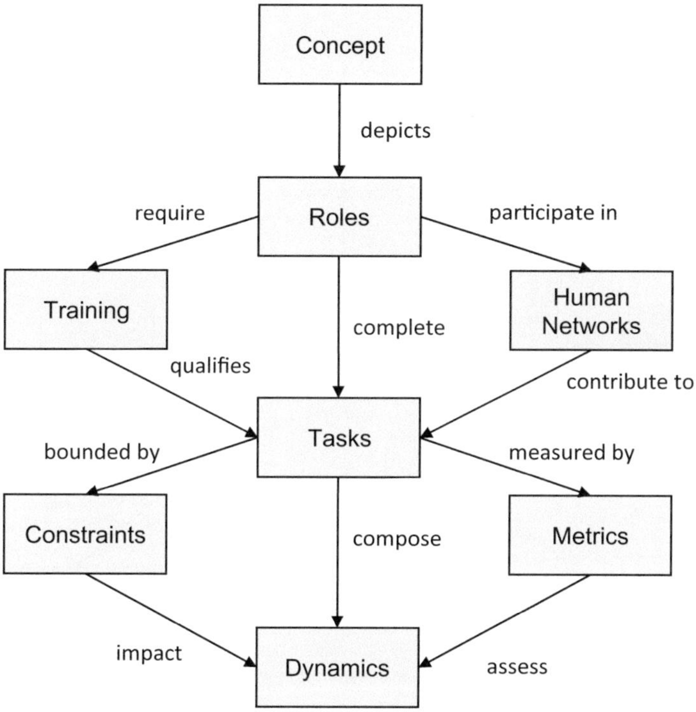

Fig. 4.3 Human View inter-relationships

be assigned to individual personnel. Additionally, the System Interfaces associated with the Tasks are captured. The Roles uses the organizations and positions and lines of authority indicated in the Organizational Chart to determine the appropriate job functions to define as Roles. Human Networks reference the Operational Nodes, which cluster Operational Activities by system, functionality or location, as well as the required Information Exchanges among the Roles. The Information Exchange identifies needed provider and consumers of data or information. The Systems Measures Matrix captures measures applicable to systems, and the Systems Technology Forecast includes predictions about trends in technology that may impact the current architecture.

The integration of the Human Viewpoint with the rest of the architecture framework provides a venue for system centric planning and development decisions. Together it delivers a set of fully integrated models that can be used to inform and influence system development. The Human Viewpoint provides a connection from the engineering community to the manpower, personnel, training, and human factors communities. It ensures that the human component has visibility as part of the system design process and provides a mechanism to highlight the impacts and required human changes resulting from system design decisions.

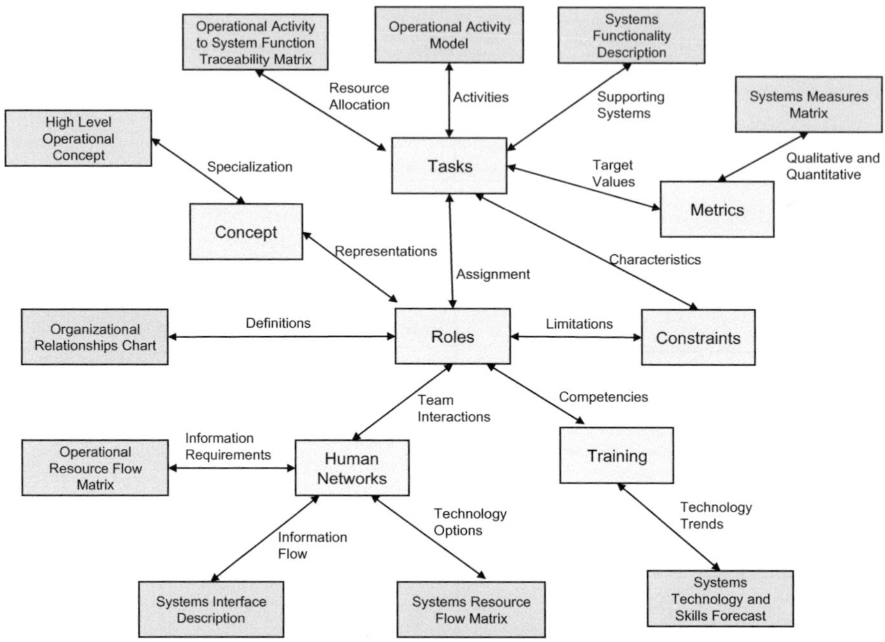

Fig. 4.4 Integrated Human Views with operational and system views

4.7 Example

A socio-technical analysis is concerned with the fit between the technology and the humans that interact with it. An example of a socio-technical analysis, based on the Human Views, is a key thread analysis. A key thread analysis examines a sequence of tasks in order to recognize performance indicators and potential risks. This provides a projection of how a given sequence of tasks will perform under different circumstances, and the implications of changes to both the human and/or technology on the process outcomes. For example, Fig. 4.1 represents a sequence of human-centered tasks from the Commander's Daily Update process described in Chap. 3. Figure 4.1 shows the task process, "Create Assigned Slides", which is one of several sub processes of the overall task process (Handley and Heacox 2005). The key thread follows the process from start to finish, identifying nodes as either human or technology supported. By using information for each node stored in the surrounding architectural products, those nodes that may impact the process outcomes can be identified and further investigated with a node analysis.

A node analysis centers on a task that has conditions that influences the choice of paths or outcomes in the work process. The analysis highlights the lack of robustness of the socio-technical system at that point and emphasizes the shifts in reliance between technology and people. Since the Human Viewpoint models capture the relationships across the socio-technical boundary, it can suggest alternatives that

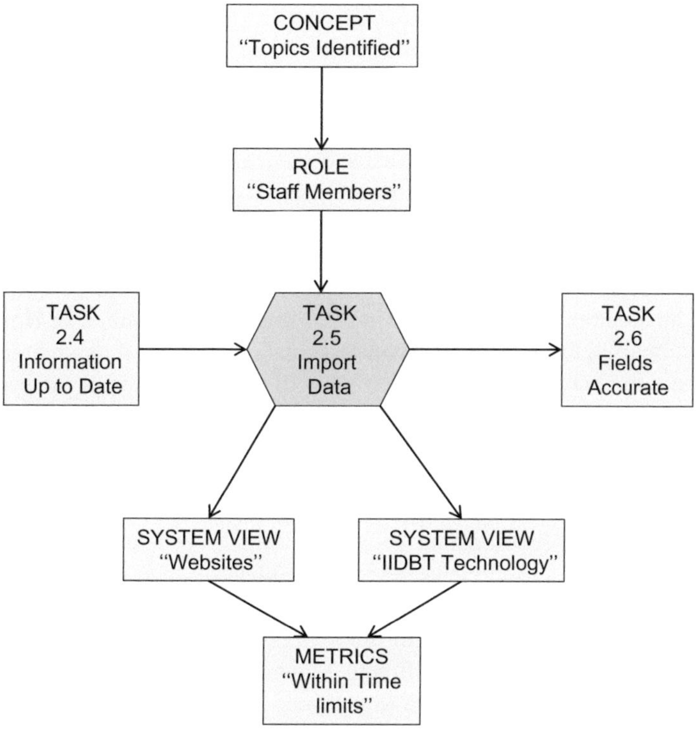

Fig. 4.5 The commander's daily update brief node analysis

might help mitigate the risk and reduce the impact (Handley 2013). An example of a node analysis of a technology assisted node is shown in Fig. 4.5. The node "Import Data", part of the key thread shown Fig. 4.1, is expanded by including the information captured in the neighboring architecture products. The items of interest for this node, as shown in Fig. 4.4, are the "Identified Topics" (from HV Concept), the assigned role (from HV Role), and the technology (from the System Views). As shown, there are options for the use of technology. In order to maintain the timeliness of this work process, it may be appropriate to use the Integrated Interactive Data Briefing Tool (IIDBT), an automated data gathering process using Web services that pulls data directly from authoritative sources. This may increase the likelihood that accurate information will be provided in a timely manner. Since the Human Views capture the relationships across the socio-technical boundary, it can suggest alternatives to mitigate risk and reduce the impact of potential events.

4.8 Summary

The Human Viewpoint was developed by a multinational committee to address the need to provide a separate viewpoint to augment current system architecture frameworks. The Human Views were designed to organize human information and identify the human requirements and parameters of a social-technical system. They provide a comprehensive representation of human capabilities by capturing sets of human information that can be used to inform and influence system design and development. The Human Viewpoint provides models which classify the human operator activities, tasks, and information exchanges required to support the system. The Human View models can be used to collect and organize social and technical parameters in order to understand the way that humans interact with other elements of the system. The resulting socio-technical analysis describes how the people, technology, and work process come together as a comprehensive system and identifies both social and technical limitations.

Currently none of the DoDAF defined viewpoints focus explicitly on collecting and organizing human-focused data. By augmenting the system architecture description with the human-focused views, a more complete set of system data is provided for development and analysis. The Human Viewpoint is fully integrated with the rest of the architecture framework in order to perform analyses and assess design impacts on the total system. The Human Views can be used to help identify and design the human requirements of a system and can be used to show the effect of high workload, poor training, and inadequate communications on system outcomes. The Human Viewpoint supports early inclusion of human considerations that can then be included in the system design to enhance human performance throughout the system lifecycle.

References

Cummings T, Worley C (1997) Organizational development and change. South-Western College Publishing, Cincinnati, OH

DoDAF (2004) DoD architecture framework version 1.0, Deskbook. Department of Defense Architecture Framework Working Group, Washington, DC

DoDAF (2010) The DoDAF architecture framework version 2.0, U.S. DoD Deputy Chief Information Officer. http://dodcio.defense.gov/Portals/0/Documents/DODAF/DoDAF_v2-02_web.pdf. Accessed 1 Nov 2018

Handley H (2013) Using architecture models to design adaptive socio-technical systems. Presented at the 2013 conference on complex adaptive systems, Baltimore, MD

Handley H, Heacox N (2005) DoDAF products for the commander's daily update brief process, baseline version. Technical report, Pacific Science & Engineering, San Diego, CA

Handley H, Smillie R (2008) Architecture framework human view: the NATO approach. Syst Eng 11(2):156–164

NATO (2010) Human systems integration for network centric warfare. Technical report, North Atlantic Treaty Organization Research Technology Organization, TR-HFM-155, Brussels, Belgium

Chapter 5
Realizing a Human Viewpoint

Abstract This chapter details the overall system architecting process that aligns with the Department of Defense Architecture Framework. It also describes in detail the Human Viewpoint methodology. This five stage methodology includes understanding the context, collecting the data, designing the models and performing analyses. It also describes the Fit for Purpose concept and its use in rendering Human Views to address specific stakeholder concerns at the last stage. The methodology is aligned with a sequenced development of the individual views and results in a complete Human Viewpoint.

Keywords Architecting process · Human Viewpoint methodology · Fit for Purpose views

5.1 Introduction

System architecting entails data identification, collection, analysis, and presentation. While an architecture framework provides a taxonomy for organizing the data, it does not prescribe an architecting process. The objective of a system architecture development is to identify and collect system data at a level of detail sufficient to address stakeholder concerns, perform trade-off analysis among competing priorities and provide the baseline configuration for further system development. Since each system architecture has unique complexities, the resulting architecture description should be developed in a way that meets the defined conditions for the system. The Human Viewpoint methodology focuses on collecting and organizing human focused data, identifying the important relationships between the data elements, and rendering views of the data to provide models of the socio-technical system. These Fit for Purpose models can be created using visual modeling representations, similar to other architecture viewpoints, and the resulting Human Views should directly address stakeholder socio-technical concerns.

H. A. H. Handley, *The Human Viewpoint for System Architectures*, Topics in Safety, Risk, Reliability and Quality 35, https://doi.org/10.1007/978-3-030-11629-3_5

5.2 System Architecting Process

Since every system is unique, there is not a standardized approach that can be consistently applied to create an architecture description. However, there are suggested methods to help guide the process. DoDAF version 1.0 offered a high-level methodology for approaching an architecture development. It described what needed to be done, but not how to do it. This generalized process consisted of the six steps shown in Table 5.1.

At the beginning of the architecting effort, it is important to determine as specifically as possible the purpose of the architecture, i.e., the issues the architecture is intended to explore and the questions the architecture is expected to help answer. This clarity will make the architecture development effort more efficient and improve the utility of the resulting architecture. The purpose determines how wide the scope needs to be, which characteristics need to be captured, and what timeframes need to be considered (Levis and Wagenhals 2000). Architecture developments can be undertaken simply to document an existing architecture or capture requirements for a proposed system, however these types of efforts also need a defined context and scope in order to set the architecture boundaries.

Once the context and scope has been defined, the prospective content of the architecture can be determined as well as the appropriate level of detail to be captured. If relevant elements are omitted, the architecture may not be useful; if unnecessary elements are included, the architecture may be confusing with details that are irrelevant to the important issues. The architecture framework chosen to organize the data will help guide the data collection based on the viewpoints and view templates provided. The architecture data is collected and the chosen views populated based on the guidance provided for that system; however, the architecture itself does not provide conclusions or answers to stakeholder questions. Additional analyses must be performed with the architecture data, models and/or simulations. Depending on

Table 5.1 DoDAF version 1.0 architecting process (DoDAF 2004)

Steps	Description
Step 1	Determine the intended use of the architecture—determine the issue or question the architecture is intended to address
Step 2	Determine the architecture's scope, context, environment—items to be considered include the appropriate level of detail to be captured and the operational context
Step 3	Determine the architecture characteristics to capture—identify what entities and relationships are required to satisfy the purpose of the architecture
Step 4	Determine which architecture views and products should be built—it may not be necessary to build the complete set of architecture views and supporting products
Step 5	Build the requisite products—the diagrams should be consistent and properly interrelated
Step 6	Use the architecture for its intended purpose—the architecture does not itself provide conclusions or answers

Table 5.2 DoDAF version 2.0 architecting process (DoDAF 2010)

Steps	Description
Step 1	Determine the intended use of the architecture—defines the purpose and intended use of the architecture, including Fit for Purpose approaches
Step 2	Determine the scope of the architecture—defines the context and level of detail required for the architectural content
Step 3	Determine the data required to support the architecture development—identifies the required data entities and attributes to support the purpose of the architecture and the Fit for Purpose query
Step 4	Collect, organize, correlate, and store the architectural data—collects and organizes data to use for model presentation and decision-making purposes
Step 5	Conduct analyses in support of architecture objectives—conducts analyses that support the Fit for Purpose implementation and support decision maker objectives
Step 6	Document results in accordance with decision-maker needs—creates views based on both pre-define architecture models and custom Fit for Purpose view to provide meaningful presentations for decision-makers

the determined purpose of the architecture, individual views can be chosen that represent the elements and relationships of interest. The final architecture description should provide information to support decision making relevant to the architected system and facilitate communication among the stakeholders.

As shown by the high-level process captured in Table 5.1, the focus of the architecting activity for DoDAF version 1.0 was the completion of the DoDAF prescribed products, rather than actively using them for analysis and decision making. With the release of DoDAF version 2.0, the architecting philosophy transformed to focus more on supporting stakeholder decisions and deemphasized the completion of compliant products. This focus is illustrated by the emphasis on the determination of data and data collection at the beginning of the process, followed by data analysis, leaving the rendering of the models that contain decision focused information as the final step. The updated DoDAF version 2.0 architecting process is shown in Table 5.2.

5.3 Human Viewpoint Methodology

Creating a Human Viewpoint is a sub-set of the overall architecting process. Both a high-level method, similar to the generalized DoDAF method, is provided to guide the overall development of the viewpoint, as well as a more detailed process to develop and integrate the individual views. The Human Viewpoint methodology provides a process to describe the human system and capture it in a set of models to augment the architecture description. It provides a sequence of iterative activities that help structure the stakeholder queries in order to identify the level of information granularity, the criterion for populating the views, and the form and format of the models. The methodology to realize a Human Viewpoint consists of five steps: Context, Data,

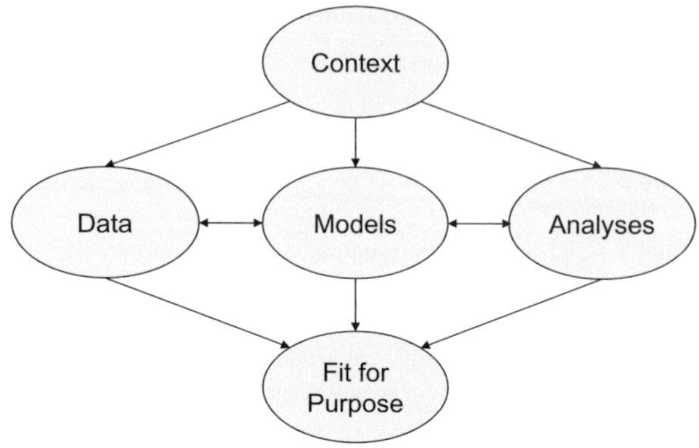

Fig. 5.1 The Human Viewpoint methodology

Models, Analysis, and Fit for Purpose (Handley and Knapp 2014). This is shown in Fig. 5.1.

Context. The context provides a high-level diagram that captures the purpose of the Human Viewpoint effort and is used to identify the scope of human focused data pertinent to the area of stakeholder concern. The context provides the overall framework for the remainder of the Human Viewpoint development by clustering high level variables into the different Human Views and identifying the relationships between them. It creates a data map that provides the overall guidance to complete the Human Viewpoint based on pertinent entities and relationships for the different Human View models.

Data. While the concept provides guidance on the categories of data to be collected, this stage focuses on identifying, collecting and organizing specific types of socio-technical system data. The data is captured in tables that include the relevant attributes of each of the elements and provides a repository of human focused data for the domain of interest. Linkages between data are identified and cross tables may be created that identify independent and dependent variables for analysis. This stage provides the context specific human data for each of the Human View models.

Models. This stage renders visual models that illustrate the important relationships between the data elements that impact the socio-technical system design. Views are created by populating model templates with the human focused data collected in the previous step. The models capture the relationships of interests that help focus on the area of concern. The models are representations similar to other architecture viewpoint models, as well as specially designed tables or diagrams that are chosen at this stage to best represent the decision data.

Analysis. This stage involves performing simulations or calculations based on different use cases to provide analytic data to support the design decisions consistent with the context. The relationships identified in the Human View analysis can be used to vary conditions such as role to task assignment and skill level to evaluate the impact of human constraints and limitations on overall system performance. The analysis stage uses models, algorithms or other tools to evaluate alternatives, answer specific questions or suggest solutions to address the purpose of the Human Viewpoint effort.

Fit for Purpose Views. The Fit for Purpose stage produces custom views that include the results of analyses to support stakeholder decisions. The models, tables and diagrams created in the model stage are annotated with the results of the analyses performed in the analysis stage and used to communicate with decision makers. In this way, the resulting Human Views that are provided to augment the architecture description directly address stakeholder concerns.

The Human Viewpoint methodology shown in Fig. 5.1 is iterative. As the initial sets of data are collected, rendered as models, and simulated or analyzed, the results are included back as expanded data sets to be included with future rendering of the models. The concept may be updated as the simulation results indicate that other data or use cases need to be included in the analysis. The final outcome of the process is Fit for Purpose Human Views, i.e., a set of Human View models with context specific data and analysis outcomes that can be used to evaluate different architecture implementations to ameliorate stakeholder concerns.

5.4 Designing the Individual Human Views

The Human Viewpoint collects human focused data to provide a basis for socio-technical system analysis. While some of this data may be included in other architecture viewpoints, the Human Viewpoint pulls it together in one integrated viewpoint to describe how humans fit in and interact with systems. The Human Viewpoint methodology produces a set of custom views during the system architecting phase to focus on particular areas of stakeholder concern.

The design of the individual the Human Views is incorporated within the overall Human Viewpoint methodology. At the context stage, high-level representations of the types of data to be included in each view is determined. At the data stage, repositories to collect and organized the data are created for each view. At the model stage, representations are determined for each view that best communicate the information to the stakeholders. At the analysis stage, information from multiple views is combined to provide the simulations or analytics required to address the stakeholder concerns. Finally, the models are rendered to provide the information required for decision making.

To assist system architects in completing a Human Viewpoint, a sequence of development for the complete set of Human View has been created (Handley and Kandemir 2013). The views have been divided into stages that naturally flow from the Human Viewpoint methodology. The first stage is initiated by describing the socio-technical system context. From this, the concept view is developed that describes the interaction of humans with the operational environment and system components.

The second stage focuses on a set of views, the Tasks, Roles and Training. Tasks describe the human activities, usually by more fully decomposing higher level functions. Roles represent job functions or task groupings. Training requirements are determined based on anticipated knowledge, skills, and ability requirements from the mapping between roles and tasks.

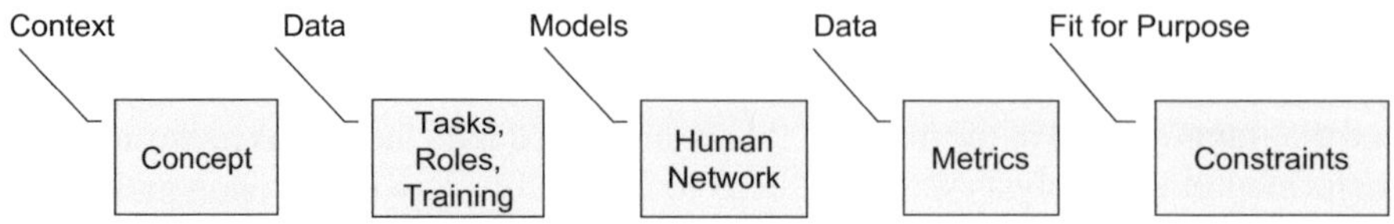

Fig. 5.2 The sequence of Human Views

The third stage focuses on human interactions and develops a Human Network model, usually represented as a work process describing the interactions of the roles completing tasks to support a higher-level task or use case. This view is created at the models stage of the Human Viewpoint methodology as it draws together data collected from multiple views.

The fourth stage focuses on the Metrics view, which is often fully developed during the analysis step of the Human Viewpoint methodology. Metrics representing human performance criteria are used to evaluate findings from different analyses on the capabilities and performance of the socio-technical system.

The fifth stage focuses on Constraints, as information provided to decision-makers must be interpreted in context of current or future human or system limitations. This sequencing of the Human Views is shown in Fig. 5.2.

5.5 Example

An example of the type of data captured in each of the Human Views is given in this section. The example data is from a communication system that supports voice and data services without the need for a fixed infrastructure; this allows the use of voice and data communications while mobile in remote regions (Handley et al. 2015). The system is installed on select military vehicles and is used to extend tactical radio networks for geographically separated elements blocked by terrain features.

For this system, the stakeholder concerns center on the current configuration of the vehicle crew, specifically if the current crew member assigned to the communication system can adequately operate the equipment. Because of the configuration of the existing vehicle, the new system will be installed with access only from the right side back seat of the four-person vehicle. This means that the system must be operated by the existing crew member that sits in this seat. The Human Viewpoint can be used to collect the data, perform analyses, and the render models to evaluate the impact on the vehicle crew due to a redistribution of responsibilities to accommodate the new system and determine if an alternative personnel type would be a better match to operate the new equipment.

Stage 1: The *Concept View* describes the interaction of humans with the operational environment and system components based on high level scenarios or use-cases. For the communication system example, the concept should articulate the high-level objectives that require the use of the equipment, as well as details on the vehicle that restrict the use of the equipment to certain personnel types. The Concept determines the data requirements for the rest of the views and appropriately scopes the analysis to the area of interest.

Stage 2. The high-level objectives articulated in the Concept provide the starting point for determining the data for the *Tasks View*. Tasks capture the set of key activities that are performed by the crew member within the context of interest. For the communication system example, the focus is on the additional tasking requirements for the crew member in the vehicle seat responsible for the new equipment. Once the tasks are defined, the *Roles View* captures the job duties for each crew member. Roles can be defined by clustering related activities and mapped to an appropriate person that assumes responsibility for the tasks. The addition of new tasks redefines the role of the equipment operator—and it may impact other crew members of the vehicle as well as tasks are redistributed. This redistribution may also impact the required skills for each defined role. For this example, the fit of personnel to the roles will be determined by identifying the required knowledge, skills and abilities and captured in *the Training View.*

Stage 3. The *Human Network* captures the interaction between human operators as part of the information exchanges required to support and complete task processes. Understanding the relationships between roles facilitates the sharing of information or assisting other roles in their duties. For the communication system example, assistance for operating the equipment resides outside of the vehicle and is available by request. The Human Network can provide additional information about the availability of additional resources, such as personnel with communication system expertise, to assist the equipment operator.

Stage 4. The *Metrics View* captures performance standards required for the human operators. Metrics can provide feedback on the ability of the candidate crew members to perform adequately on the communication equipment tasks, Analyses can predict operator success rates and workload values to evaluate the human impact on system performance. The analysis provided on the communication equipment will focus on the ability of the current crewmember to perform the equipment tasks within the expected standards, along with the trade-off analysis of replacing the current crew member with a different personnel type.

Stage 5. The *Constraints View* captures limitations that impact the ability of personnel to assume different roles or complete assigned tasks. For the communication equipment example, the constraints will state the manpower and personnel availability of the different types of crew members that are candidates for the assignment. Additionally, other human factors constraints, such as the operating environment (i.e., vehicle size restrictions), work cycle (i.e., watch standing requirements), and system interface restrictions may further limit personnel options.

5.6 Summary

An architecture framework defines a common approach for the collection, presentation, and integration of architecture data. It is intended to provide a consistent method for the development of architecture descriptions. While a framework does not provide detailed guidance on how to complete an architecture description, generalized processes have been designed associated with the DoDAF framework. The Human Viewpoint methodology is a sub process within system architecting with the goal of capturing human focused data to support stakeholder discussions and decision making. The Human Viewpoint methodology collects and categorizes socio-technical system data to provide information and analyses on the human component of a system. The resulting Human Views address specific stakeholder questions and display specific sets of data on the impacts to the current workforce.

The Human Views help visualize human roles and workflows, and account for human capabilities and limits based on task demands, training, and environment factors. They support interface design by grouping tasks into roles and aligning interface functionality with role requirements. Describing the human work processes provides requirements for task sharing between operators and identifies the information elements necessary for roles to complete tasks. Role and task data, along with scenarios that provide environmental data, can also be used to develop user profile information describing how the operator will need to interact with the system; the constraints provide additional information that shape the interface requirements. For the example communication system, the resulting Fit for Purpose views created will support stakeholder concerns on the ability of the current crew member to operate the new equipment. Alternatives to the current crew member who may perform better can also be evaluated, however these suggestions are subject to the limitations provided by the manpower and personnel constraints.

References

DoDAF (2004) DoD architecture framework version 1.0, Deskbook. Department of Defense Architecture Framework Working Group, Washington, DC

DoDAF (2010) The DoDAF architecture framework version 2.0, U.S. DoD Deputy Chief Information Officer. http://dodcio.defensegov/Portals/0/Documents/DODAF/DoDAF_v2-02_web.pdf. Accessed 1 Nov 2018

Handley H, Kandemir C (2013) Human view considerations of the intelligence crew for the multi-intelligence platform long endurance. Technical report, Old Dominion University, Norfolk VA

Handley H, Knapp B (2014) Where are the people? The human view approach for architecting and acquisition. Def Acquis Res J 21(4)(71):851–874

Handley H, Kandemir C, Dario E (2015) Warrior information network-tactical human view models & analysis. Technical report, Old Dominion University, Norfolk, VA

Levis A, Wagenhals L (2000) C4ISR architectures: I. Developing a process for C4ISR architecture design. Syst Eng 3(4):225–247

Chapter 6
Stage 1: Context Development and the Concept View

Abstract This chapter introduces the notion of "as-is" and "to-be" architectures, and the use of architecture views to plan the transition. It describes the initial stage of the Human Viewpoint methodology, the context development. It emphasizes the use of a data map at this stage to identify the required data and relationships to be captured in the Human Views. It also introduces the first view to be developed, the Concept view, which provides the guiding parameters or use case for the remainder of the Human Viewpoint development.

Keywords Context development · Concept view · Data map

6.1 Introduction

Usually the hardest part of completing an architecture description is getting started. Initial architecting activities focus on determining the scope of the effort, understanding the system components and boundaries, and identifying what elements are important to include that are pertinent to the stakeholder interests. The Human Viewpoint development emphasizes a Fit for Purpose approach, implying the views should be developed with the end in mind. This ensures that the models created as part of the architecture description are relevant to address stakeholder concerns. The first stage in the Human Viewpoint methodology is the development of the context. This identifies what the elements of the socio-technical system are and helps determine what types of data should be collected, the appropriate models to develop, and the analyses to perform. The Concept view provides information about the socio-technical system in relation to the environment, operational demands, and technical components. This view helps communicate the overall purpose of the human focused architecture and identifies the specific types of data to collect to focus the views appropriately for the Fit for Purpose development.

H. A. H. Handley, *The Human Viewpoint for System Architectures*, Topics in Safety, Risk, Reliability and Quality 35, https://doi.org/10.1007/978-3-030-11629-3_6

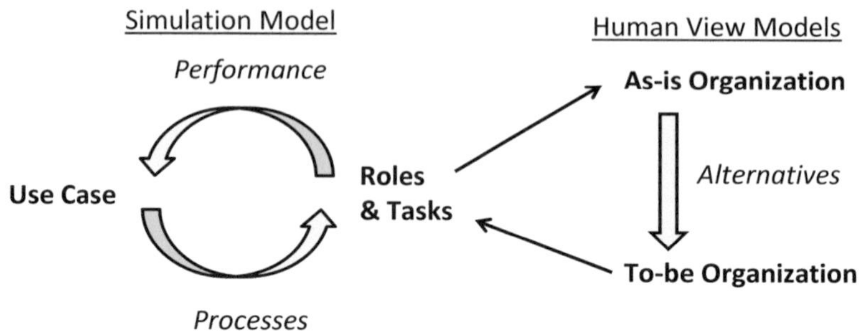

Fig. 6.1 As-Is and To-Be design process

6.2 As-Is and To-Be Views

In many cases, the Human Viewpoint models are used to capture the condition of the current socio-technical system, or the "as-is" state, and then used to design alternative, "to-be" states. For example, the Human Views can be used to create a responsibility matrix by mapping the tasks to the responsible roles. By further extending the responsibility matrix with different sets of constraints, i.e., the skills or experience levels required for each task, alternative crew assignments can be identified. These configurations can be simulated to evaluate both the workload of the crew members assuming different roles, and also the performance of the process under different scenarios or use cases. Once satisfactory levels of performance have been achieved, the design options for the to-be socio-technical system based on the crew member assignments can be finalized. This process is shown in Fig. 6.1.

The Human Viewpoint can facilitate the design of alternative operator and task arrangements through the analysis of simulation data. The operator requirements can be determined by evaluating the roles, tasks, and work process with different sets of constraints; this information can be provided back to the simulation model to evaluate its effectiveness in the operational environment. The Human View architecture can then be used to capture the human system requirements of the "to-be" organizational design. As personnel are reassigned to new tasks, based on the constraints of required skills and workload thresholds, the new responsibilities can be captured in the Human View models.

6.3 Human Viewpoint Methodology—Context Development

Architecture definition activities center on determining the intended use of an architecture, thus developing the context is the first stage of the Human Viewpoint methodology, see Fig. 6.2. The context provides a high-level diagram that describes the

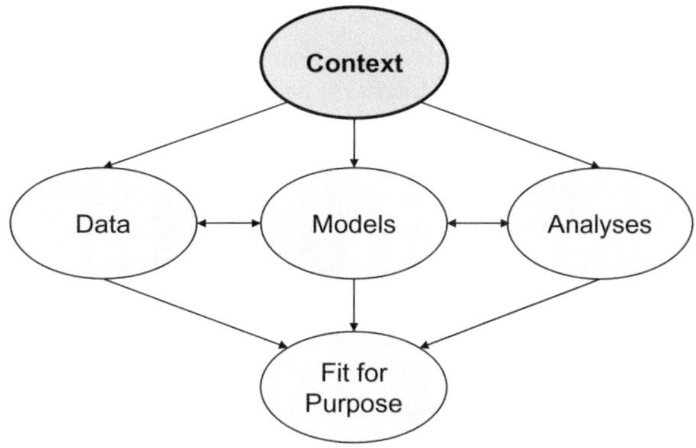

Fig. 6.2 Human Viewpoint methodology Stage 1. Context

purpose of the Human Viewpoint effort. The context provides the overall framework for the remainder of the socio-technical architecture development by clustering high level variables into the different Human Views and identifying the relationships between them.

The context is based on the operational concept for a system that describes the interaction of humans with the system and the environment. This includes the characteristics of the human users, the need for operator decision points, as well as major functions and information flows. The resulting context development provides a common starting point for both the system architects and the human system engineers without being overly prescriptive regarding the final socio-technical system design.

The Human Viewpoint methodology context development leverages concepts from soft systems design (Checkland 2001). Soft systems design is a way to describe a system with a visual representation of the major functions, activities, and information flows. Soft systems methodology translates the operational concept to a pictorial representation. The goal is to capture the operational concept in a way that can be used to reason about the socio-technical system and plan the Human Viewpoint development. A rich picture is a soft system design high level expression of the operational concept through the use of elements and relationships, represented with cartoon or stick figures. Figure 6.3 is an example of a rich picture.

While rich pictures are a good initial effort to understand the operational concept of the socio-technical system, the Human Viewpoint methodology requires a more detailed assessment of the specific data required to address the stated stakeholder concerns. The Human Viewpoint context requires a data map, i.e., a hybrid of a systemigram and an entity relationship diagram (Handley and Knapp 2014). Systemigrams present stakeholders with a visual representation based on a written description of the system under consideration using standard conventions (Boardman and Sauser 2008). Entity relationship diagrams model data relationships: An entity is the representation of some data that is to be depicted in the diagram and a relationship is an association that exists between two entities (Chen 1976). The resulting data map identifies the types of data that need to be collected in each of the views and the meaningful relationships between them. A template of a data map is

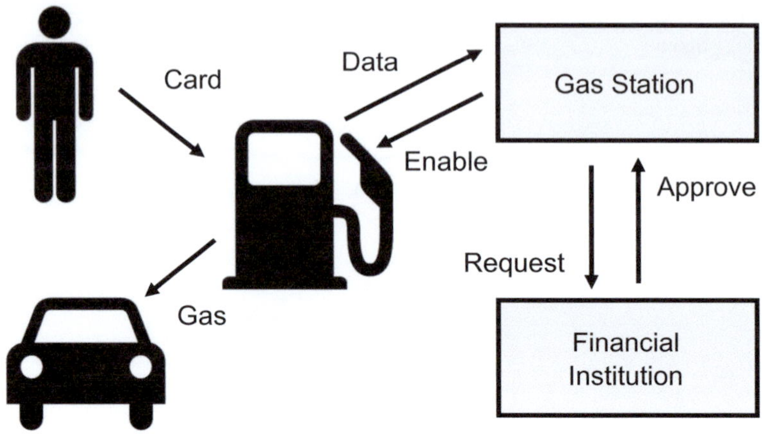

Fig. 6.3 Rich picture example

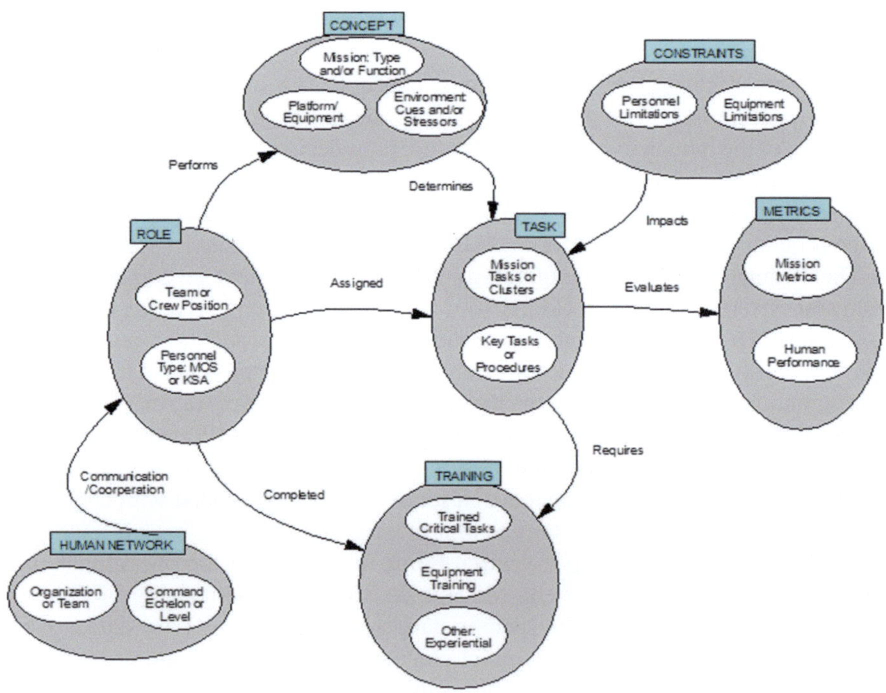

Fig. 6.4 Context data map template

shown in Fig. 6.4. As shown in the figure, each of the Human Views is shown as a large node, with typical data types identified in each of the internal nodes. High-level relationships between the views are identified by the arrows.

The data map is customized for each application, depending on the purpose of the architecture development and the specific stakeholder questions. One issue encountered at the beginning of an architectural development is how to identify the relevant data entities and their attributes as well as the appropriate models to develop. This is especially difficult for Fit for Purpose architectures as they are triggered by stakeholder questions, and the questions may not be at a level of granularity that matches with usual architectural data. Thus, the data map is an essential first step in the Human Viewpoint methodology to establish the framework for providing a socio-technical architecture that supports specific stakeholder concerns.

Since there are multiple ways to configure and populate each Human View depending on the data available and the analysis being performed, the data map provides guidance for the content specific data to be captured for each of models. It also identifies the linkages between them which provides the foundation for the Human Viewpoint analysis step by classifying the independent and dependent variables. These relationships can be further explored to identify potential "to-be" versions of the socio-technical system.

6.4 Human Views—Concept

The data map created as part of the context development strives to capture the operational concept, with a focus on the human component, in a way to support stakeholder discussions about the system. The Human View Concept captures the parameters of a specific use case or scenario that will be used to configure the Human Viewpoint analysis. While the context provides guidance for the overall project development, the Concept view provides the level of detail for the specific stakeholder evaluation. It is the first view usually developed, see Fig. 6.5.

The original definition of the Concept view was closely related to the high-level operational graphic included within the overall system architecture description. It differed by trying to include aspects of the human component in the socio-technical system, usually as cartoon figures associated with a system or capability (Handley and Smillie 2008). Its purpose was to provide a visualization of the human component in the context of the system functions. However, it was often difficult to communicate through a high-level graphic the specific types of concerns that drove the need for a socio-technical architecture analysis. Additionally, with the use of visual modeling tools and the need to export data to simulation tools, it was determined that the

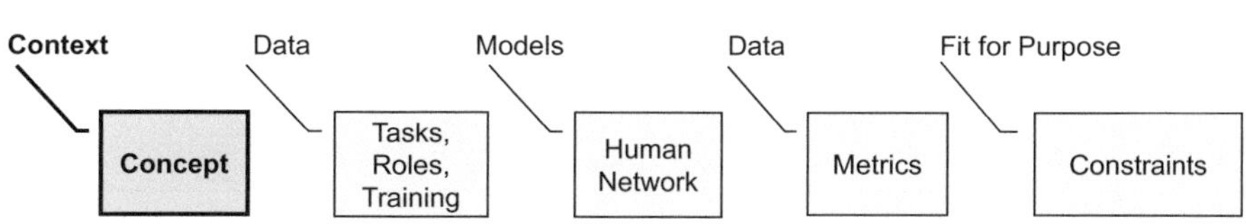

Fig. 6.5 The concept view in the Human Views sequence

concept would be better served by identifying the use case or scenario data that can be used to bound the human focused analysis.

The resulting Concept view helps identify the area of concern for the socio-technical system by providing information on the operational environment based on high level scenarios or use-cases. It can identify changes in conditions that impact the human component and form the foundation for the stakeholder concerns. The concept view may limit the interaction of humans with system components in order to focus the area of analysis.

6.5 Example

The previous chapter introduced an example communication system that supports voice and data services without the need for a fixed infrastructure; this allows the use of voice and data communications while mobile in remote regions (Handley et al. 2015). The system is installed on select vehicles and is used to extend tactical radio networks for geographically separated elements blocked by terrain features. The first stage of a Human Viewpoint development for this system requires the creation of a data map to help capture the specific data of interest for each of the Human Views. The template of Fig. 6.4 can be customized for this example as shown in Fig. 6.6.

As shown in the data map of Fig. 6.6, the Concept view captures information on the vehicle type, as the configuration of the vehicle determines how many crew

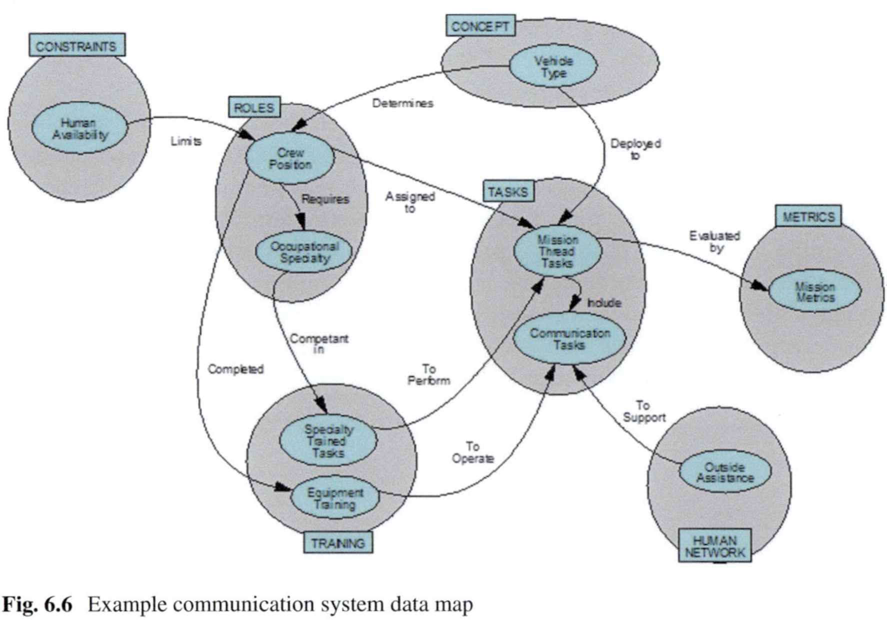

Fig. 6.6 Example communication system data map

Table 6.1 Concept view data for communication system example

Crew	Seat 1	Seat 2	Seat 3	Seat 4
Current	Driver	Commander	Infantryman	Gunner
Alternative	Driver	Commander	Signal specialist	Gunner

members are required and the roles of those crew members. Crew members must have the appropriate qualification for the assigned position, as shown in the Roles view, based on their occupational specialty. The Training view captures the specialty training received, as well as any equipment specific training required to perform the communication tasks. The Tasks view captures the mission tasks pertinent to the vehicle crew and the communication equipment. The Metrics view captures parameter used to evaluate the mission performance and the Constraints view indicates limitations to personnel assignment based on availability. The Human Network view indicates the ability of communication equipment assistance from specialists outside of the vehicle.

The Concept view provides the environment for the human system based on high level scenarios or use-cases. As indicated in the data map, the Concept view for this example should capture data about the vehicle that contains the communication equipment, as the type of vehicle determines the characteristics of the crew. The Concept view also identifies the changes in conditions that is motivating the need for a Human Viewpoint analysis. The data for the Concept view for the communication equipment is shown in Table 6.1.

For the communication equipment example, the stakeholder concerns focus on determining the best operator for the new equipment, i.e., a different type of crew member may be better qualified to complete the tasks that require the new communication equipment. The concept bounds the data appropriate to the area of concern, in this case the crew member assigned to the third seat. The Concept view captures the current configuration of the vehicle crew and a proposed alternative, a Signal Specialist, replacing the Infantryman as the operator of the communication equipment.

6.6 Summary

This chapter describes the first stage of the Human Viewpoint methodology. The context provides the description necessary to plan and initiate the Human Viewpoint architecting effort. It identifies important data elements and the relationships between them pertinent to the stakeholder area of concern, in the context of the individual Human Views. Since there are multiple ways to configure and populate each individual view, the data map provide guidance for the content specific data to be collected in each model.

The Concept view bounds the area of concern for the socio-technical system analysis by providing information on the operational environment based on high

Table 6.2 Communication system example Human Views data

Views	Content
Concept	Vehicle crew positions
Tasks	
Roles	
Training	
Human network	
Metrics	
Constraints	

level scenarios or use-cases. It can identify changes in conditions that impact the human component and forms the foundation for the analyses to address stakeholder concerns. The data for the Concept view is the first step in populating the set of Human Views, as shown in Table 6.2.

References

Boardman J, Sauser B (2008) Systems thinking: coping with 21st century problems. CRC Press/Taylor & Francis Group, Boca Raton, FL

Checkland P (2001) Soft systems methodology. In Rosenhead J, Mingers J (eds) Rational analysis for a problematic world revisited. Wiley, Chichester

Chen P (1976) The entity-relationship model—toward a unified view of data. ACM Trans Database Syst 1(1):9–36

Handley H, Kandemir C, Dario E (2015) Warrior information network-tactical human view models & analysis. Technical report, Old Dominion University, Norfolk, VA

Handley H, Knapp B (2014) Where are the people? The human view approach for architecting and acquisition. Def Acquis Res J 21(4)(71):851–874

Handley H, Smillie R (2008) Architecture framework human view: the NATO approach. Syst Eng 11(2):156–164

Chapter 7
Stage 2: Collecting Data for Tasks, Roles and Training Views

Abstract This chapter focuses on the second stage of the Human Viewpoint methodology and the multiple views that are completed at this stage. The data collection stage of the methodology leverages the information from the data map to create tables of data appropriate for each of the identified Human Views. The next set of views to be designed are the Tasks, Roles and Training views. These views capture the information on the human specific activities, the job descriptions, and the relationships between them that form the basis for the remainder of the Human Viewpoint.

Keywords Data collection · Responsibility matrix · Tasks view · Roles view · Training view

7.1 Introduction

The previous chapter described the first stage of the Human Viewpoint methodology, developing the context. The context helps guide the rest of the viewpoint development by identifying the pertinent data of interest for each of the Human Views. The next stage in this process is to identify the specific sources of data for each of the views and to begin to collect the necessary data in tables or other repositories. While the first stage created the Concept view, the data stage focuses on the Roles, Tasks and Training views. These Human Views are at the heart of a socio-technical analysis as they capture the work that the human component is responsible for, along with the job descriptions and required training. Most of the questions answered by a socio-technical analysis center on if the right person is assigned to the right job with the required skills.

7.2 Human Viewpoint Methodology—Data Collection

The data map created in the previous step identifies the socio-technical system data pertinent to the stakeholder question. The data map also provides guidance for the content specific data to be captured for each of the Human Views. This is an important

H. A. H. Handley, *The Human Viewpoint for System Architectures*, Topics in Safety, Risk, Reliability and Quality 35, https://doi.org/10.1007/978-3-030-11629-3_7

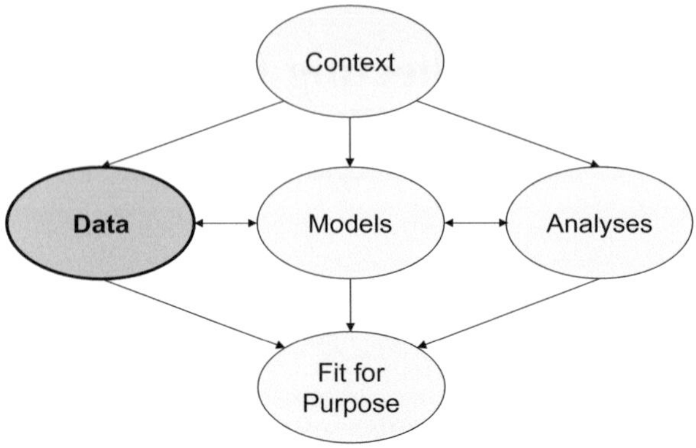

Fig. 7.1 Human Viewpoint methodology Stage 2: Data

step as there are multiple ways to configure and populate each view depending on the data available and the analysis being performed. The next stage, data collection, refers to identifying data sources and collecting the data for each view as defined in the data map, see Fig. 7.1. While the high-level content of the data was defined in the context stage, the data may be both refined and expanded as appropriate to the area of inquiry as indicated by the Concept view.

Generally, the data collection starts by creating tables for each of the Human Views, with columns for each of the high-level variables identified in the data map. As the problem becomes more fully developed, and sources of information are identified, the data elements may become more detailed with additional attributes added. These data tables provide an intermediary between the high-level definitions described by the data map and the specific elements that will be rendered in the visual models. Additionally, most architecting tools require the user to first create data tables before choosing the data elements and rendering model diagrams—the tables created at this stage can often be ingested directly into an architecting tool environment. Cross tables are also often created at this stage to capture the relationships between the different elements, as also presented in the data map. These cross tables identify the independent and dependent variables that will be used later during the analysis stage to evaluate the impacts of different human configurations. At the outcome of this stage, the tables of initial sets of data that can be used to populate the Human Views have been created.

7.3 Human Views—Tasks, Roles and Training

The three views described in this section are the launching point for most of the socio-technical analyses performed with the Human Viewpoint. Together these views collect information on the tasks and task requirements, roles and associated skills, and the gaps identified between what the humans are being ask to do and what they

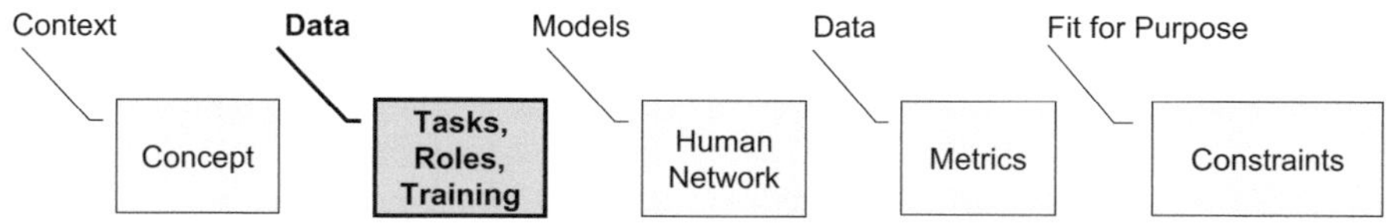

Fig. 7.2 The task, roles and training views in the Human Views sequence

are capable of performing (Handley and Smillie 2008). The relationship of these views within the Human Views development is shown in Fig. 7.2.

The Tasks view describes the human-specific activities required by humans interacting or participating in a socio-technical system. (The term "task" refers to a piece of work that can be assigned to a person). Often the Tasks view provides a decomposition of functions that are represented in other parts of the system architecture that have been allocated to human operators or users. These are the human-specific tasks that support the overall operational and system activities. The task decompositions are useful for a socio-technical analysis in terms of identifying task requirements, user skills, and performance parameters. Describing the tasks in terms of various criteria, including knowledge, skills and abilities (KSA), allows a mapping to the available personnel and determines the need for additional training. Alternatively, tasks may also be represented in task network form (or a task graph) that indicates sequences of tasks and their interdependencies, i.e., a work process. This approach allows for the inclusion of systems interfaces in the process and facilitates the allocation of specific tasks to automation.

The Roles view describes the jobs that have been defined for the socio-technical system. Roles represent the set of responsibilities that can be assigned to a person. The Roles view provides a description of the roles and their attributes in such a way that it allows them to be paired against the task descriptions to define task assignments; it usually includes the competencies required to assume that role, often in terms of KSAs. Roles are defined based on responsibilities for a particular set of tasks and associated equipment or system interfaces. The interrelationships between roles provides the basis of the organizational arrangement, inferring an authority and accountability structure.

The Training view is used to identify gaps in KSAs to meet the job requirements for a particular role assignment. It provides a detailed accounting of different types of training in order to identify where remediation is required by new task assignments, equipment or system interfaces. It can suggest the additional training to provide personnel the task competencies needed to meet job requirements. It supports personnel planning by identifying the availability of individuals who already have the necessary competencies. The Training view can be further developed as needed to address the availability and suitability of existing training resources, the assignment of alternative personnel to mitigate the impact on training, as well as specific competencies to be trained.

A complete mapping of responsible roles to individual tasks results in a matrix that allows the exploration of different responsibility assignments. A template for

Role ID: Skill List / Task ID: Requirements	Task 1.0 sk1.0	Task 1.1 sk1.1	Task 1.2 sk1.2	Task 2.0 sk2.0	Task 3.0 sk3.0	Task 3.1 sk3.1	Task 3.1.1 sk3.1.1	Task 3.1.2 sk3.1.2	Task 3.2 sk3.2	Task 3.3 sk3.3	Task 4.0 sk4.0
Role 1: sk1.0, sk3.0, sk3.1.2	X				X			X			
Role 2: sk1.1, sk3.2		X							X		
Role 3: sk2.0, sk4.0			X							X	
Role 4: sk2.0, sk4.0				X							X
Role 5: sk3.2				X							X

Fig. 7.3 Role × task matrix template

a Role × Task Matrix is shown in Fig. 7.3. The matrix can be used to highlight mismatches between role qualifications and task requirements, identify potentially over-loaded roles, and to assign primary and secondary responsibilities. From the matrix, the necessary set of KSAs for a role across all assigned tasks can be determined. Based on the task summary, the roles can be mapped onto existing crew positions and training gaps identified. The role-task matrix may also include references to equipment or system interfaces necessary to complete the task, which can impact alternative crew assignments as well as training.

7.4 Example

An example of the data stage of the Human Viewpoint methodology, specifically populating the data tables for the Tasks, Roles and Training views, is provided by continuing the communication equipment example from the previous chapters. Recall the system is used to extend tactical radio networks and is installed on select vehicles (Handley et al. 2015). The area of stakeholder concern, and the focus of the socio-technical analysis, is on evaluating whether the crew member in the vehicle seat that has access to the communication system can adequately operate the new equipment. The Concept view bounded the data appropriate to the area of concern, as shown in Table 7.1. The Concept identified the current configuration of the vehicle crew with the Infantryman assigned responsibility to operate the communication equipment, as well as a proposed alternative, a Signal Specialist, assigned to that crew position.

The next views to be developed for this example are the Tasks, Roles and Training views. The Tasks view provides the tasks that will be assigned to the communication

Table 7.1 Concept view data for the communication system example

Crew	Seat 1	Seat 2	Seat 3	Seat 4
Current	Driver	Commander	Infantryman	Gunner
Alternative	Driver	Commander	Signal specialist	Gunner

Table 7.2 Tasks view data for the communication system example

Communication equipment tasks	Task priority
Configure and call using softphone	1
Configure and utilize net radio	1
Monitor and utilize applications	2
Troubleshoot connectivity	3

Table 7.3 Sequence of subtasks for "configure and call using softphone"

Step	Subtask
1	System user activates the softphone application
2	System user dials system gateway operator
3	System user waits for connection acknowledgement
4	System user transfers control to the caller

equipment operator. These tasks can initially be at described at a high level, with further iterations to decompose the tasks to a level where skills can be assessed as well as workload and other human performance metrics. For this example, the tasks that the crew member will need to perform as part of the new communication equipment is included in the Tasks view (Handley 2017). The Tasks view data is shown in Table 7.2.

An example of a further decomposition of the high-level task "Configure and Call using Softphone" is shown in Table 7.3. A softphone is a program that emulates standard telephone calls over a network using a computer, rather than using a traditional telephone. The task data at this level can be used to identify specific steps in the work process where the operator interactions with the equipment occur.

The Tasks view captures the set of tasks that are performed as part of the socio-technical system. For this example, the data captures the tasks that require the user to interact with the equipment to facilitate crew communication across the network. The data collected for the Tasks view is consistent with the purpose determined for the viewpoint development—determining the appropriate soldier to assign to operate the communication equipment. The task data collected would be tailored differently for different stakeholder queries. For example, if the concern focused on the ability of the soldier to balance multiple tasks during a particular mission event, the tasks that are occurring simultaneously would be collected in order to evaluate the workload imposed on the soldier.

The Roles view captures the requirements and responsibilities of different personnel. Roles are usually defined initially by leveraging existing role descriptions;

Table 7.4 Roles view data for the communication system example

Seat three	Communication equipment operator	
	Current	Proposed
Occupation	Infantryman	Signal specialist
Responsibility	Basic equipment operator	Communications
Communications usage	General user	Trained user
Experience level	2–3 years	4–6 years

the roles are then revised or new roles created based on different task assignments. For example, most organizations have general personnel categories that include both level of expertise, i.e. entry level, mid-grade, experienced, as well as functional responsibility, i.e. Engineer, Human Resources, etc. Roles can capture one or both of these elements as well as additional descriptors that characterize the desired job function.

For the communication system example, the roles are defined based on known military occupation specialties as well as experience levels. As shown in Table 7.4, the communication equipment operator role will be evaluated for both the existing Infantryman as well as a suggested alternative, a Signal Specialist. The Roles view provides initial personnel descriptors that include attributes that can be linked to detailed soldier databases, such as test results, skill levels, years of experience, and associated training.

While for this example the data collection focuses on the role of operating the communication equipment, personnel usually assume multiple roles in completing their different task assignments. A more extensive role data collection would characterize the different responsibilities of the vehicle crew personnel with all of the assigned tasks based on the multiple types of equipment that they operate.

The purpose of the Training view is to identify the skills that the roles currently have in order to evaluate gaps that may be problematic if assigned to certain tasks. The Training view can focus on the training each role receives as part of their normal career development, or training that will be received as part of the new equipment deployment plan. For this example, two sets of training data can be collected: the skills that each occupational specialty receives as part of their basic training and the equipment specific training each role receives as part of the communication equipment deployment plan. The combination of these provides an overview of the competencies of the alternative personnel under considerations to be assigned to the equipment operator role.

Table 7.5 provides the Training view data from the basic task training received by each of the occupational specialties under consideration (Handley 2017). As shown in the table, specific gaps in training were noted for the Infantryman in three of the four tasks listed. Additionally, only 10% of the Infantryman's overall training is directly related to communication tasks, whereas 100% of the Signal Specialist training is communication related.

Table 7.5 Training view data from basic training

Task	Training data	
Occupational specialty	Infantryman	Signal specialist
Configure and call using softphone	No	Yes
Configure and utilize net radio	Yes	Yes
Monitor and utilize applications	No	Yes
Troubleshoot connectivity	No	Yes
Percentage of course topics (%)	10	100

Table 7.6 Training view data from equipment training

Equipment training topics	Infantryman	Signal specialist
Install, configure, and maintain basic equipment, i.e., Softphone	x	x
Application training	x	x
Troubleshoot basic equipment		x
Deployment and configuration of system infrastructure		x
Length of training	1 week	5 weeks

However, personnel will be given new equipment training prior to being deployed with the communication equipment. The type and length of training is shown in Table 7.6 (Handley 2017). Differences in the content of the training are adapted to the knowledge base of the different occupational specialties.

The data collected for the Tasks, Roles and Training views can be cross referenced to determine the "gap" in personnel training and suggest additional training that is required to operate and maintain the equipment. Table 7.7 provides a summary of the training received by each candidate role aligned with the task list. The results highlight the main area of concern—troubleshooting the system when it loses connectivity.

This stage of the Human Viewpoint methodology, the collection of appropriate data, supports the stated purpose for the communication equipment example of considering stakeholder concerns of the correct personnel for the vehicle crew. Because the equipment is in a fixed location inside the vehicle, it is not an option to simply reassign the tasks to a more qualified crew member. In this case, the role location is fixed and the person at that location must be capable to operate the equipment. While the current Infantryman assigned to the position does not have

Table 7.7 Cross table of task, role and training data

Task	Training	
Role	Infantryman	Signal specialist
Configure and call using softphone	Equipment specific	Basic and equipment specific
Configure and utilize net radio	Basic and equipment specific	Basic and equipment specific
Monitor and utilize applications	Equipment specific	Basic and equipment specific
Troubleshoot connectivity	None	Basic and equipment specific

specific communication knowledge, using a Signal Soldier as an alternative may result in assigning an over qualified and underutilized person. Identifying the correct role candidate is the desired outcome of the analysis.

7.5 Summary

The second stage of the Human Viewpoint methodology focuses on collecting and organizing data in individual tables based on the elements identified in the data map. Cross tables are also created that identify the important relationships between the data elements. The data collected for the three Human Views highlighted in this chapter, Tasks, Roles and Training, identify the human tasks and procedures, defines roles by groups of tasks or job functions, and determines potential gaps in training. The data captured in these tables will provide the basis for rendering models in the next chapter. The communication equipment example was further developed in this chapter, and the content data for the Tasks, Roles and Training views was collected, as shown in Table 7.8.

Table 7.8 Communication system example Human Views data

Views	Content
Concept	Vehicle crew positions
Tasks	Communication equipment tasks
Roles	Infantryman and signal specialist descriptions
Training	Basic and equipment specific training
Human network	
Metrics	
Constraints	

References

Handley H (2017) Defining human capability from soldier and task data. Paper presented at the human factors and ergonomics society 2017 annual meeting, Austin, TX

Handley H, Smillie R (2008) Architecture framework human view: the NATO approach. Syst Eng 11(2):156–164

Handley H, Kandemir C, Dario E (2015) Warrior information network-tactical human view models & analysis. Technical report, Old Dominion University, Norfolk, VA

Chapter 8
Stage 3: Rendering Models and the Human Network View

Abstract This chapter introduces the use of the System Modeling Language, especially in the context of Model Based System Engineering. It describes the use of standard templates to render models of the human focused data. The third stage of the Human Viewpoint methodology is presented with a mapping of the applicable model templates for each of the Human Views. The Human Network view is described in detail, illustrating the ability to draw data elements and their inter-relationships from other views and incorporating them in a single, compound view.

Keywords Model based system engineering · System modeling language · Human networks view

8.1 Introduction

This chapter continues the Human Viewpoint methodology by introducing the model stage. The Human Views adopt the DoDAF version 2.0 definitions of models and views: a model is a template for a visual representation of a certain set of data; when the model is populated with data specific to a system it becomes a view of that system. The model templates employed for the Human Views are based on the System Modeling Language (SysML). The Human View described in detail in this chapter is the Human Network view. This view captures the interactions of roles completing tasks and can include information sharing and other forms of collaboration. The view can also capture sequencing and coordination of tasks; these temporal dependencies between tasks is an important component of the evaluation of socio-technical system performance.

H. A. H. Handley, *The Human Viewpoint for System Architectures*, Topics in Safety, Risk, Reliability and Quality 35, https://doi.org/10.1007/978-3-030-11629-3_8

8.2 Human Viewpoint Methodology—Rendering Models

The third stage of the Human Viewpoint methodology uses the data collected in the previous stage to render views, or populated model templates, that describe the configurations and relationships of the socio-technical system, see Fig. 8.1. The design of the models to be rendered is based on the types of data collected and an understanding of the purpose of the Human Viewpoint. Models can be any representation that describes the data in such a way to share knowledge.

Using models helps architects visualize the data and relationships collected in the previous step, and share this information with other stakeholders in a more understandable format. System engineering has embraced the use of models throughout the system engineering process. Model Based System Engineering (MBSE) is the formalized application of modeling to support system requirements, design, analysis, and validation activities beginning in the conceptual design phase and continuing throughout development and later life cycle phases (INCOSE 2015). Using models to realize the architectural description reduces ambiguity, supports a more collaborative development environment, and facilitates communication with stakeholders. Modeling ensures a seamless transition to the follow-on system engineering activities from the descriptive architecture models to the prescriptive engineering design solution.

Using MBSE for the Human Views has the same benefits as for architecting in general. Models provide a common language that increases understanding of both the technical factors as well as the human factors for the developing system. Because it is not possible to build a meaningful single model that encompasses the whole system, the set of Human Views address different aspects of the socio-technical system and provide a representation of human concerns understandable by stakeholders from different technical backgrounds.

The Systems Modelling Language (SysML) is a general-purpose visual modelling language for systems based on the Unified Modeling Language (UML) which was developed for software engineering (Friedenthal et al., 2015). SysML has been

Fig. 8.1 Human Viewpoint methodology Stage 3: models

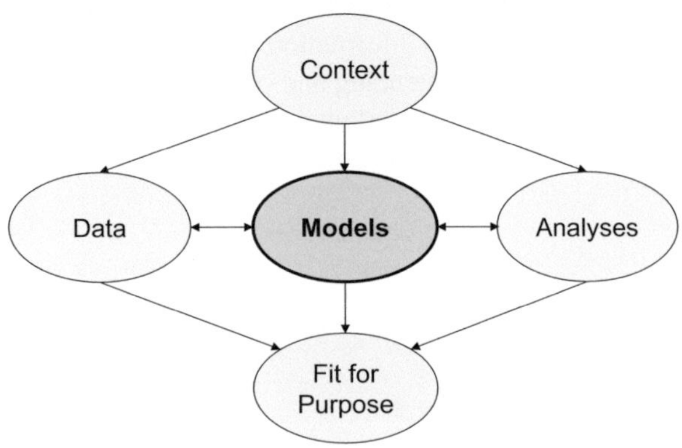

Table 8.1 SysML diagram types (OMG 2015)

Diagram type	Diagram	Description
Structure	Block	Provides a general-purpose capability to represent system elements and their relationships
	Internal block	Captures the internal properties and connections of system elements
	Package	Provides the ability to organize models into groups or viewpoints
Behavior	Use case	Describes at a high level how external actors interact with the system
	Activity	Represents the internal inputs, outputs and control of activities
	Sequence	Captures the sequential exchanges of data between system elements
	State machine	Describes the states of a system and the transitions between the states in response to triggering events
Parametric		Contains properties that support engineering analysis and govern the operation of the system
Requirements		Captures the capabilities or conditions that must be satisfied

adapted to support the system engineering community and has become the modeling language of choice for MBSE (Hause and Moore 2006). The data collected for the Human Views can be used to populate the SysML pre-defined, standardized representations to provide the human-focused models. By using the standard SysML formats, the Human Views are in the same format as the overall system architecting efforts. Thus, SysML becomes a "common language" for human system engineers to integrate with other system engineering efforts by leveraging the standardized SysML constructs and diagrams.

SysML provides nine different templates that are used to capture the elements and relationships of a system. The SysML models fall into four categories: structural, behavioral, parametric, and requirements. The structure models describe the physical and logical organization of a system, while the behavioral models are used to describe what the system does and how it operates. The parametric and requirements diagrams were specifically added for use by system engineers. The parametric model is used to represent system parameter values to quantify system operation and performance. The requirement model captures the requirements and constraints that drive system development. The nine diagrams are described in Table 8.1 (OMG 2015).

Appropriate SysML diagrams can be identified as Human View model templates and used to capture the human focused architecture information. While the traditional Human Views focus on one area of concern, such as tasks or roles, the SysML approach emphasizes capturing both the entities and their relationships to other elements of the system. A summary of the mapping from the traditional tabular data collected for the Human View to potential SysML visualizations is shown in Table 8.2.

Table 8.2 Human Views mapped to SysML diagram types

	Block	Use case	Activity	Sequence	State machine	Parametric	Requirements
Concept		X		X	X		
Tasks	X		X				
Roles	X		X	X			
Training							X
Human Network	X			X			
Metrics	X					X	
Constraints	X						X

When translating from tabular data to a SysML representation, both the context of the problem and the relationships among the views are considered in order to choose the diagram that best communicates the purpose of the Human View data.

The mapping shown in Table 8.2 was developed by investigating the properties of the individual SysML diagrams as potential Human View model templates. For example, the Roles model data can be used as the "roles" property in a SysML Block Definition diagram. The same role data can also be used as "lifelines" in a SysML Sequence diagram and as "activity partitions" in SysML Activity diagrams. Another example is the Metrics model data. The Metrics model is used to define performance parameters and standards associated with the socio-technical system. For instance, workload values defined in this Human View model can be translated as "equations" or "constraints" in a SysML Block Definition diagram. The same constraints can be used in the SysML Parametric diagram.

SysML provides the templates that can be populated with data to render the Human View models; what aspects of the system are modeled and which templates and data are chosen depends on the socio-technical system and the stakeholder concerns. As shown in Table 8.2 different templates can be chosen based on the type of data collected and the context of the analysis. However, in order to facilitate its use to develop the Human Viewpoint, a subset of SysML models have been assigned as Human View templates (Handley and Amissah 2015). Table 8.3 shows the typical SysML diagram to use to render the Human Views and the content data. As noted, most diagrams require data from multiple Human View data tables to support the relationships necessary to complete the diagram. This is a due to the fact that the Human Views were designed to focus on individual categories of data, while most of the SysML diagrams also highlight the relationships between the data.

Table 8.3 SysML templates selected for the Human Viewpoint models

Human View	SysML diagram	Diagram content
Concept	Use case diagram	Environment, user scenarios
Tasks	Activity diagram	Tasks, inputs, outputs
Roles	Block diagram	Role, relationships
Training	Block diagram	Role and/or task and KSAs
Human network	Sequence diagram	Roles, tasks, information exchanges
Metrics	Parametric diagram	Parameters that govern operations
Constraints	Requirements diagram	Limitations that must be considered

8.3 Human Views—Human Network

The Human Network model focuses on the interaction of the human elements of the system: how the tasks are distributed among the roles and what interactions between the roles are required to complete the tasks. The Human Network is a compound view as it usually includes data from both the Roles and Tasks views and adds the interconnections required to share data or other communication sequences. The Human Networks captures the human to human communication patterns that occur as a result of task collaborations. It can also identify the information required between humans and systems, determining the points where information is exchanged resulting in a human-system interface. The Human Network view in the sequence of Human Views development is shown in Fig. 8.2.

The Human Network view includes the inter-dependencies between different tasks, the information demands to perform specific tasks, and the tools required to accomplish a task. For socio-technical systems, the collaboration requirements between distributed roles and the resulting communication patterns are of particular importance. The Human Network view focuses on capturing the parameters and variables that characterize the human communication processes and can provide

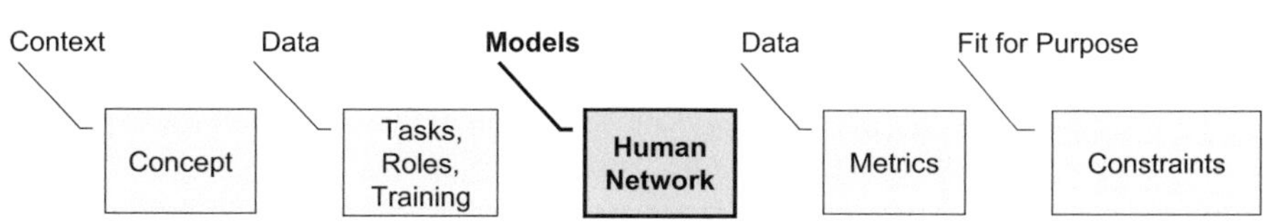

Fig. 8.2 The human networks view in the Human Views sequence

the necessary data for a time-based simulation model for a dynamic evaluation of the socio-technical system. By providing a mechanism for capturing the required data to characterize the human interactions, the Human Network is an important view for designing human centered systems and evaluating the impact of the human component on overall system performance.

The Human Network is also an important view for analysis of crews or other work teams, especially virtual teams. Virtual teams exist when tasks are distributed across a team and the team is also distributed across physical locations; the team members may also reside in different organizations. This has implications as to which types of communication technologies are used, how complex data can be shared, and how responsibilities are distributed to ensure effective communication. Understanding the relationships between roles helps support the task process by identifying the formal and informal communication channels that exist in the organizational design.

8.4 Example

An example of the Human Viewpoint model stage and the Human Network view is provided by continuing the communication equipment example from the previous chapters. Recall the Concept identified the current configuration of the vehicle crew with the Infantryman in the seat required to operate the communication equipment, as well as a proposed alternative for the position, a Signal Specialist. The data collected for the Tasks, Roles, and Training views identified the communication equipment tasks, described characteristic of the Infantryman and the Signal Specialist, and identified the basic and equipment specific training received to identify potential gaps in the ability to complete the communication equipment tasks.

At this stage, the Human Viewpoint methodology renders visual models of the data collected for the individual views. The Human Network model depicts the interactions between the roles to support the operator in completing the communication tasks. The Human Network visual model draws data from both the Tasks and Roles views to describe the required series of activities. An appropriate SysML template for this model view is the Sequence diagram; when populated with the system specific data it is rendered as the Human Network view. This is shown in Fig. 8.3.

The Human Network view shown in Fig. 8.3 represents the work process for accomplishing the softphone task. The model depicts the interrelationships of tasks, roles and information required to complete this higher-level task. The sub tasks can be modeled separately in other diagrams allowing a greater level of detail. An example of the sub task sequence to support the softphone task using a SysML Activity diagram is shown in Fig. 8.4.

The Human Network view also supports a social network approach for capturing authority or responsibility relationships between roles. For example, the Human Network view could also capture the availability of assistance for the communication equipment operator in the extended organizational network. The Human Network

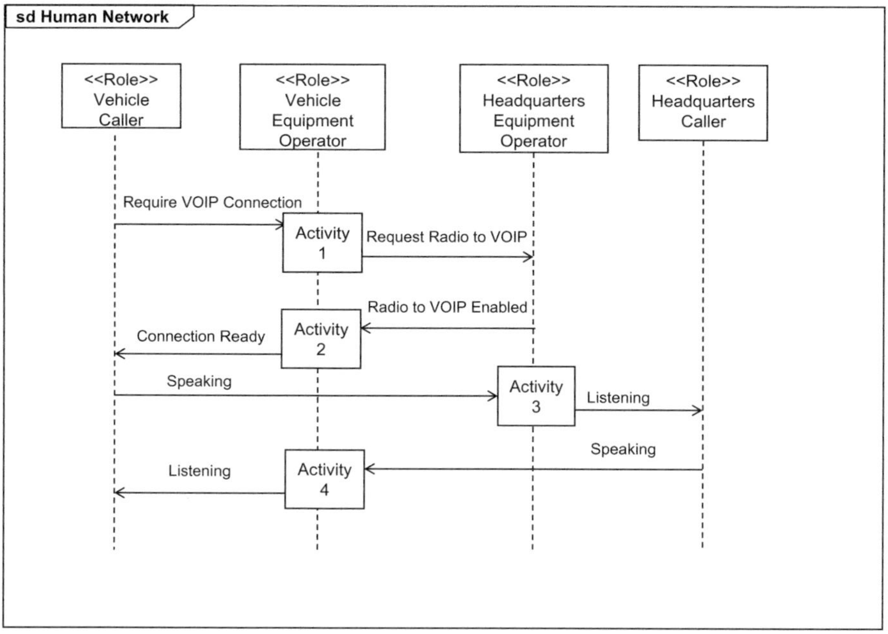

Fig. 8.3 The human network view using a SysML sequence diagram

Fig. 8.4 Activity diagram for the softphone task

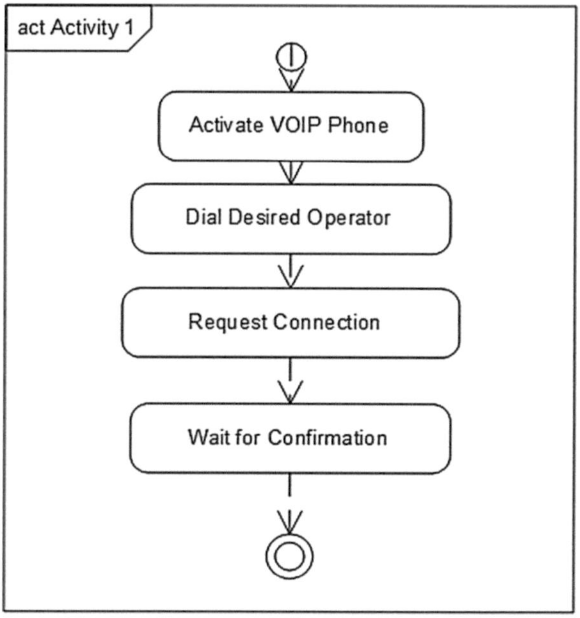

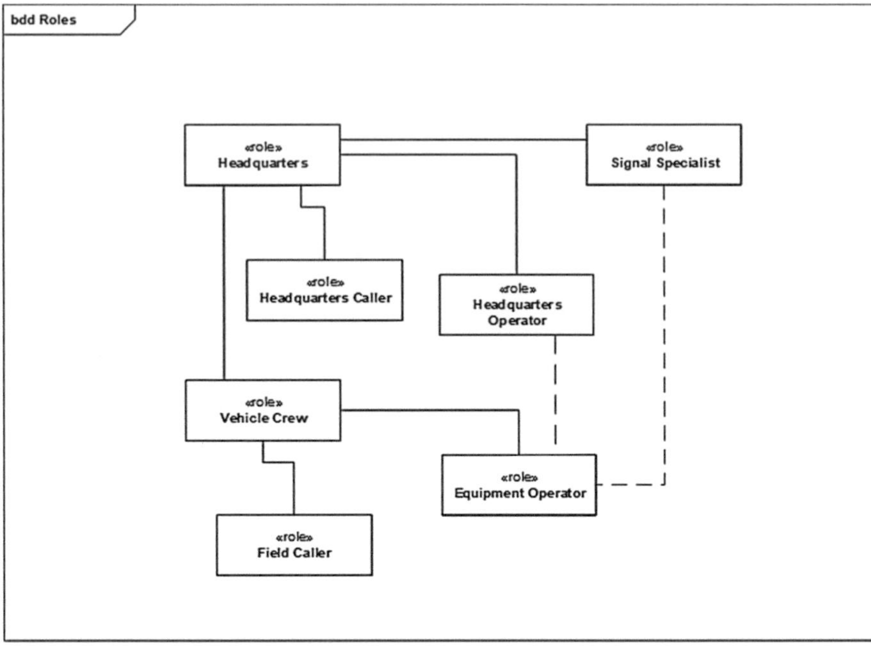

Fig. 8.5 The human network view using a SysML block diagram

view shown in Fig. 8.5 is a different rendering, using a SysML Block diagram, to describe the relationship of the communication equipment operator to nearby roles.

As shown from the examples for the Human Network view, each SysML diagram can be tailored to convey specific information about the socio-technical system; the selection of which diagrams to use depends on the information to be conveyed and the type of data available. The Human Network sequence diagram focuses on the information exchanges between tasks and roles to complete specific tasks. The Human Network block diagram focuses on the social network structure of the organization. Depending on the goals of the analysis, different templates can be chosen and additional data can be included in the rendering of the models to present different views of the socio-technical system.

8.5 Summary

MBSE focuses on capturing system design information using integrated models; it promotes the use of models to share the outcomes of the system architecting process vice the traditional document-based products. SysML is a graphical modeling language adapted to support the system engineering community; it is often used as a general-purpose visual modeling language for system engineering applications.

Table 8.4 Communication system example Human Views data

Views	Content
Concept	Vehicle crew positions
Tasks	Communication equipment tasks
Roles	Infantryman and signal specialist descriptions
Training	Basic and equipment specific training
Human network	Interactions of the communication operator with other roles
Metrics	
Constraints	

SysML can be used to create the models for a Human Viewpoint development and provides a common language to share models between architecting and engineering teams. MBSE facilitates integration, reuse and consistency of architecture data.

The Human Viewpoint methodology advocates rendering models of the human focused data using SysML diagrams. Since there are multiple ways to configure and populate each Human View, the SysML diagram type and content data are selected based on the guiding context for the Human Viewpoint development. Using SysML templates provides consistency both within the Human Viewpoint and across the larger system architecture development, thus the SysML diagrams provide an important link between the Human Viewpoint and other system architecture viewpoints.

The outcome of the rendering models step is a set of Human Views with context specific data that can be used to perform analyses to support stakeholder decision making. The Human Network view identifies information exchanges and coordination between roles completing tasks. The Human Network view is often the link to the Human Viewpoint analysis stage as it captures sequencing and other time dependencies that are necessary to evaluate the performance of the socio-technical system. The communication equipment example was further developed in this chapter, and the data collected for the Human Network view is shown in Table 8.4.

References

Friedenthal S, Moore A, Steiner R (2015) A practical guide to SysML, the systems modeling language. Elsevier, The Netherlands

Handley H, Amissah M (2015) Model based soldier decision framework: Integration of human view and IMPRINT. Technical report, Old Dominion University, Norfolk, VA

Hause M, Moore A (2006) The systems modeling language. ARTISAN Software Tools, Beaverton, OR

INCOSE (2015) Systems engineering handbook, 4th edn. The International Council on Systems Engineering, San Diego, CA

OMG (2015) Object management group, systems modeling language specification. www.omg.org/spec/SysML. Accessed 9 Sept 2018

Chapter 9
Stage 4: Performing Analyses and the Metrics View

Abstract This chapter emphasizes the use of the human-focused data to perform analyses to address stakeholder concerns. The fourth stage of the Human Viewpoint methodology advocates different analyses for both human and technical aspects of the system with the goal of improving system performance while remaining within other constraints, such as cost and schedule. It provides examples of analysis that can be performed with both single and multiple Human Views. This chapter also details the Metrics view and identifying applicable measures to evaluate the socio-technical system.

Keywords Trade-off analyses · Variance analysis · Metrics view

9.1 Introduction

This chapter continues developing the Human Viewpoint methodology by describing the analysis phase, i.e., using the data and models to answer questions about the socio-technical system. The term "analysis" can take on different meanings in the context of the stakeholder query; it can be an evaluation of the fit and/or gaps within the human focused data, or it can use analytic techniques to assess the trade-offs between different socio-technical system elements. It most cases, the analysis uses combinations of the Human View models developed to suggest solutions within the area of concern. This chapter also describes in detail the Metrics view. This model may have wide variation depending on the system under consideration, but usually identifies parameters of interest to the socio-technical system. Often the view contains two sets of data, the first concerning the human component and the second concerning the human system performance. The human focused metrics may include measures such as workload, availability or competence, while the system focused metrics may indicate minimum requirements concerning timeliness, accuracy or throughput. Taken together, the two sets of metrics can be used to evaluate the balance of the human and technology components of the system.

H. A. H. Handley, *The Human Viewpoint for System Architectures*, Topics in Safety, Risk, Reliability and Quality 35, https://doi.org/10.1007/978-3-030-11629-3_9

9.2 Human Viewpoint Methodology—Analysis

This stage of the Human Viewpoint methodology uses different analysis techniques on the human focused data and models in order to address the concerns of the socio-technical system. Figure 9.1 shows the interaction of the stages of the Human Viewpoint methodology. Analysis is a rather loose term that refers to the detailed examination of the elements and relationships as a basis for interpretation, discussion, or revision (Dictionary 2018). There are many analysis methods available and the choice depends on the data available and the question to be answered.

Some analyses are performed on a single view or model of the Human Viewpoint; individual views capture different aspects of the socio-technical system and can be evaluated for feasibility, consistency, or missing data. For example, task sequences can be examined for the required flow of data and to identify bottlenecks or long delays. Different sequences can be evaluated to identify sets of roles that may need to interact. A variance analysis can examine the details of a single task that has conditions that influence the choice of sequential tasks; this analysis identifies the sets of limitations that impact the task and estimates the impact of changing these conditions. The outcome of any of these analyses can identify socio-technical system performance issues and highlight the sensitivity of specific tasks.

However, many of the questions of interest for the socio-technical system can only be addressed by understanding the relationships between elements captured across multiple views. For example, generating a cross product of the tasks with the assigned roles can identify the different skill sets associated with each role in the context of the requirements of each task. Varying the role to task mapping can highlight alternative task assignments that can influence task performance, while also illustrating the impact the reassignments may have on the task load of other roles. An assessment of the workload demand of roles completing specific task sequences can be projected by using a technique such as the Task Analysis—Workload method (Bierbaum and Hamilton 1990). Using a comprehensive task decomposition of the tasks allocated to the different roles, the estimated the visual, cognitive, auditory and psychomotor

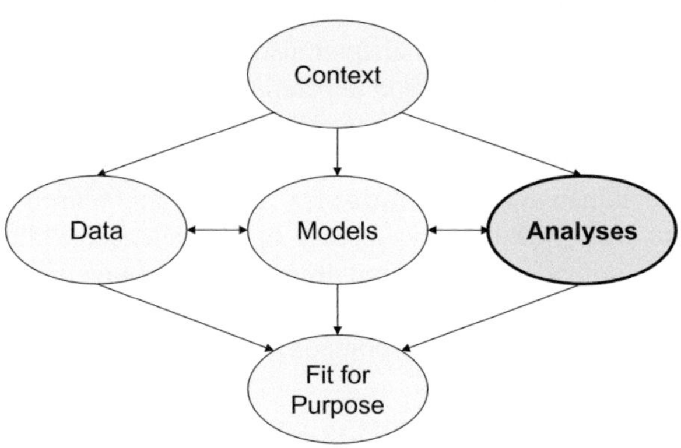

Fig. 9.1 Human Viewpoint methodology Stage 4: analysis

(VCAP) workload demands can be assessed on a seven-point scale. Workload levels can significantly influence the performance of operators, and a rebalancing of the tasks may be required to ameliorate overloaded conditions. These types of analysis identify the strength and weakness of different role and task configurations.

Additional analyses performed with combinations of Human View models can use system dynamic techniques to evaluate the balance between human and technical capabilities as one resource is increased or decreased (Krasnecky and Curry 2015). It is a way to represent and evaluate trade-offs between different entities and can provide a summary of the positive and negative effects between human and system elements. For example, increased automation may reduce overall manpower, but it may also increase the number of specialists required. Changes to one element can be traced through the Human View models to determine positive and negative impacts on the socio-technical system. A later chapter describes the use of risk-based decision analysis techniques to evaluate the likelihood of system success based on risk factors within the system.

Finally, a comprehensive simulation model can be created by integrating information from multiple Human Views. A simulation model can be used to evaluate the impact to both the human operators and on system performance when adding new tasks or technologies. For example, when mapping new tasks onto existing roles, generally one of two methods is used. If the new tasks are similar to existing tasks, they are usually mapped to the role who already has the expertise for that type of task. In this case, the existing role is now assigned more tasks. The second method is to map new tasks to existing roles that are underutilized. In this case, a better balance of workload among the roles is maintained, however some roles may need additional training to become proficient on the new tasks. An analysis performed using a simulation model can show the impact of the task assignment decisions on overall system performance. Varying skill levels can show how assigning a role with less experience may lead to a failure to complete key tasks, affecting overall system performance; likewise, performance degradation due to overloaded roles can lead to delays and dropped tasks. The simulation model can show the impact to the human elements of a system and suggest mitigations to balance the concerns of the socio-technical system (Handley and Smillie 2010). A later chapter describes in detail the use of the Improved Performance Research Integration Tool (IMPRINT) to develop simulation models based on the Human Viewpoint to provide socio-technical dynamic analyses.

9.3 Human Views—Metrics

The data map completed in the Human Viewpoint methodology context stage identifies the high-level variables that can be manipulated in the analysis phase to explore the stakeholder's area of concern. The Metrics view contains the different criteria used to evaluate the alternative configurations of the socio-technical system; the sequencing of Human Views is shown in Fig. 9.2. For example, varying the skill or experience level required for each role and evaluating the overall impact on personnel

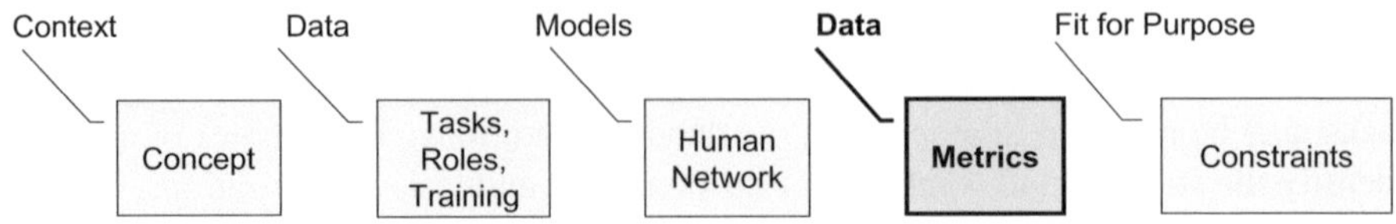

Fig. 9.2 The metrics view in the Human Views sequence

Table 9.1 Example of human focused metrics

Metrics	Description
Human limitations	
Operator workload	Component overloads, total workload, thresholds and problematic tasks
Workload density	Weighted workload, indicates high demand of a task
Organization design	
Load balance	Distribution of workload, tasks between employees
Human availability	Busy-idle time (processing time vs. monitoring, communication time)
Human performance	
Timeliness	The time delay between start and end of the task
Accuracy	The percent of correct steps completed for the task

requirements, as well as system performance, provides information to stakeholders on alternative role configurations. The Metrics view provides the criteria that both the role availability and the task performance are evaluated against to help determine an acceptable solution.

The Metrics view provides a repository for human-related measures. It may include human performance criteria, such as timeliness and accuracy, as well as human specification metrics, such as workload and availability. Within the Human Viewpoint analysis, it is important to show how the human will impact performance at the system level, such as mission success, supportability, and cost, as well as how the human component will be impacted by the system operating within the identified environment, such as personnel availability, skill demands, and training requirements. The Metrics view is used to provide parameters that can be used to evaluate the ability of the socio-technical system to perform within technical specifications and within human limitations. As with all the views, the Metrics view can be adapted for the specific Human Viewpoint development described by the context. An example of human focused metrics is shown in Table 9.1.

9.4 Example

The Human Views developed for the communication equipment example during the previous chapters focus on a specific stakeholder question: Is the current crew member in seat three of the vehicle capable of operating the new equipment or should a Signal Specialist be assigned? At the core of this specific query is the difference in domain specific knowledge that is required by the system; the Infantryman receives equipment training but does not have the in-depth communication training that a Signal Specialist receives. The cross table of Tasks, Roles and Training data can be used as the starting point for this analysis, as it provides a summary of both the basic training and the additional equipment training that both candidate roles receive; the cross table created in the data collection stage is shown in Table 9.2. As indicated in the table, the main skill gap for the Infantryman is a lack of training to support the Troubleshooting Connectivity task.

Troubleshooting involves applying logic to search for the source of a problem in order to solve it. The troubleshooting process is a systematic approach to identify and check system components until the error is found and corrected (Quick 2003). The specific skill gap due to the lack of troubleshooting training for the Infantryman can be identified by examining the underlying abilities that contribute to this task. Troubleshooting invokes abilities from the "reasoning" skill cluster (Fleishman et al. 1984). These abilities enable the soldier to adapt procedures to new situations and apply the rules to new information. The Signal Specialists basic training includes more focus on reasoning skills such as numerical analysis and problem-solving skills that support troubleshooting tasks.

While replacing the Infantryman with a Signal Soldier may be one solution, another alternative is to rely on roles outside of the vehicle crew for assistance with this task. The Human Network view provides information on the social system available to support the Infantryman operating the communication equipment; qualified Signal Specialists reside one level up in the organizational hierarchy. These soldiers are skilled in troubleshooting the communication equipment and are capable of providing virtual assistance to the vehicle crew.

Table 9.2 Cross table of task, role and training data

Task	Training	
Role	Infantryman	Signal specialist
Configure and call using softphone	Equipment specific	Basic and equipment specific
Configure and utilize net radio	Basic and equipment specific	Basic and equipment specific
Monitor and utilize applications	Equipment specific	Basic and equipment specific
Troubleshoot connectivity	None	Basic and equipment specific

Table 9.3 Metrics view data for the communication system example

Metrics	Description
Human performance	
Timeliness	Softphone calls should be initiated with 10 s
Accuracy	Softphone calls should connect on the first attempt
System performance	
Availability	The softphone should be operational 95% of the time (5% downtime for maintenance)

The Metrics view provides the performance parameters used to evaluate the acceptability of the proposed alternatives. Metrics applicable to the equipment operator include timeliness and accuracy criteria, i.e., the time delay incurred in making softphone calls and the number of attempts before the call is connected. Metrics for the technical system performance include availability, i.e. limited downtime for the communication equipment, reflecting the importance of the troubleshooting task. The Metrics data for the communication equipment is shown in Table 9.3. The data can be rendered with a SysML activity diagram including the relevant tasks, as shown in Fig. 9.3.

The potential impact to both the human performance timeliness metric and the system performance availability metric are illustrated in Fig. 9.4. When the Infantryman fails to connect the equipment after the limited number of trials (two), the operator is instructed to contact the external Signal Specialist. Depending on the availability and response time of the Signal Specialist, there is a delay added to the time required to complete the call. If there is a long delay, it may also impact the overall availability time of the communication system. The inclusion of the metrics data with the task data using a SysML Activity diagram can provide the basis for the analysis of alternatives for the socio-technical system.

9.5 Summary

The Human Views provide the data and models to perform analyses of different configurations of the socio-technical system. By leveraging the relationships between the different views, trade-off analyses can be used to evaluate the impact of alternative socio-technical system designs. For example, alternative role to task assignments that consider differences in skill sets can be traced through the Human Views to determine positive and negative impacts to task performance; these results can also be used to understand the impact of the operator on overall system performance. These types of analyses also support comparing "as-is" and "to-be" architectures in order to identify acceptable alternatives; by comparing the Human Views of an existing architecture to a future architecture, the differences can be identified and an acceptable alternative selected.

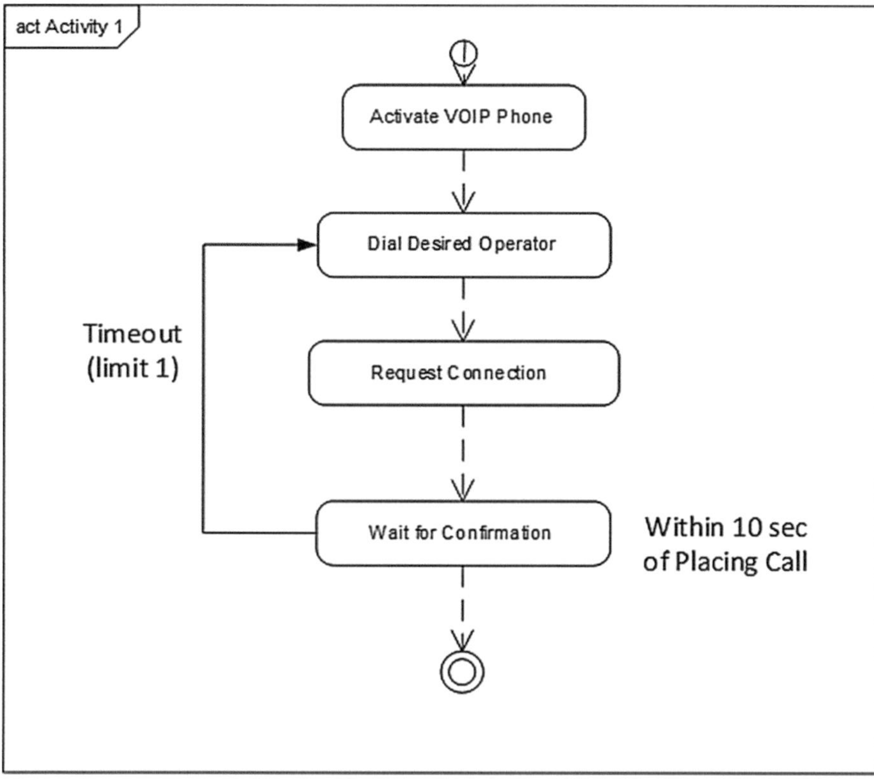

Fig. 9.3 The metrics view for the communication system example

The Metrics view provides parameters that can be used to evaluate the ability of the socio-technical system to perform within system specifications while maintaining operation within human limitations. For the example communication system, the differences in skills for the two candidate role types can be traced to the overall task process and associated performance standards. When the equipment operator requires assistance, it may impact the socio-technical system's ability to achieve its performance objectives, as captured in the Metrics view. The Metrics view for the communication equipment example captured both human and system performance measures, as indicated in Table 9.4.

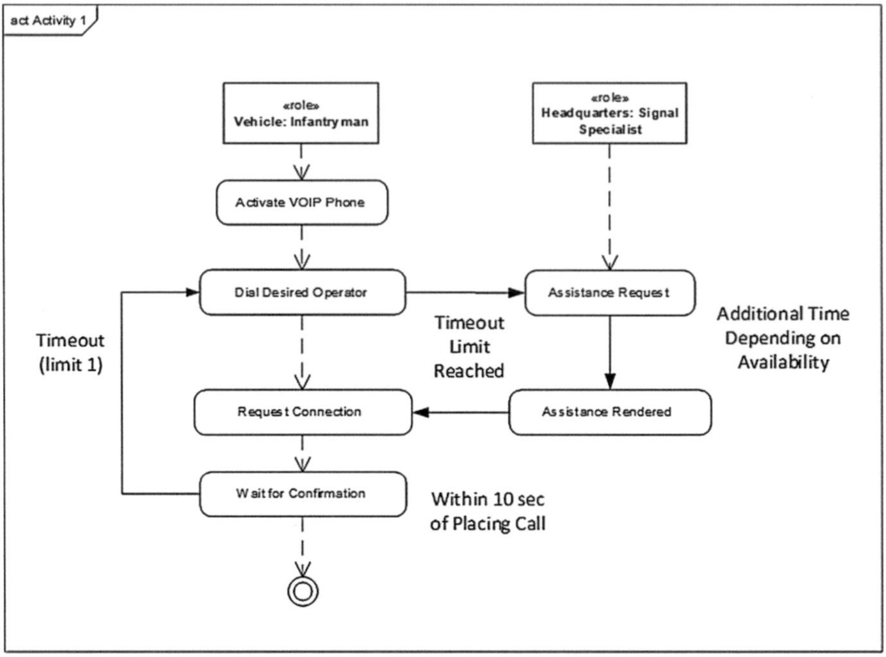

Fig. 9.4 SysML activity diagram with metrics for troubleshooting support

Table 9.4 Communication system example Human Views data

Views	Content
Concept	Vehicle crew positions
Tasks	Communication equipment tasks
Roles	Infantryman and signal specialist descriptions
Training	Basic and equipment specific training
Human network	Interactions of the communication operator with other roles
Metrics	Softphone human and system performance measures
Constraints	

References

Bierbaum C, Hamilton D (1990) Task analysis and workload prediction model. Research report, Anacapa Sciences, Inc., Santa Barbara, CA

Dictionary (2018) Google dictionary. www.google.com. Accessed 9 April 2018

Fleishman E, Quaintance M, Broedling L (1984) Taxonomies of human performance: the description of human tasks. Academic Press, Cambridge, MA

Handley H, Smillie R (2010) Human view dynamics—the NATO approach. Syst Eng 13(1):72–79

Krasnecky D, Curry M (2015) Systems engineering process. Presentation at the SSEA review, US Army Research Lab, Simulation and Technology Training Center, Orlando, FL 10–11 March 2015

Quick R (2003) Troubleshooting skills can be learned. Maint Technol 16(9)

Chapter 10
Stage 5: Fit for Purpose Outcomes and the Constraints View

Abstract This chapter describes the last stage of the Human Viewpoint methodology, combining the models and analyses to create a set of Fit for Purpose views that directly address stakeholder concerns and support strategic decision making. These custom views respond to the initial context or use-case that drove the development of the Human Views for the socio-technical system. This chapter also introduces the last Human View, the Constraints view. This view contains the limitations of the human system and may influence the interpretation of the socio-technical analysis within the realities of its implementation.

Keywords Fit for Purpose presentations · Stakeholder concerns · Constraints view

10.1 Introduction

The last stage in the Human Viewpoint methodology is to combine the data, models and analyses to produce Fit for Purpose views. "Fit for Purpose" refers to the ability to create new types of visualizations or augment existing models that are designed to directly support decision making and answer specific stakeholder queries (DoDAF 2010). These views should provide custom representations to address the questions that framed the Human Viewpoint context and focused the Human View analysis. The last Human View to be described is the Constraints view. This view provides bounding information for the Fit for Purpose presentations. Constraints capture different limitations that may impact final design decisions. It can provide, for example, restrictions on the number of people available and the percentages of people who have the expertise needed for new types of systems. The Fit for Purpose views provide the impacts of trade-off analyses embedded within the Human Views to advise on different architecture alternatives for the socio-technical system.

H. A. H. Handley, *The Human Viewpoint for System Architectures*, Topics in Safety, Risk, Reliability and Quality 35, https://doi.org/10.1007/978-3-030-11629-3_10

10.2 Human Viewpoint Methodology—Fit for Purpose Views

The last stage of the Human Viewpoint methodology is to implement Fit for Purpose views; this stage represents the culmination of the previous stages as shown in Fig. 10.1. The most important aspect of completing an architecting description is communicating the results with stakeholders. The Fit for Purpose approach includes custom views in the architecture description that particularly focus on stakeholder concerns. They can be created by including the results of analyses within the model diagrams. Fit for Purpose views are the result of the Human Views data centric approach, which collects human focused data first, presents it in Human View models second, and then tailors the models to support specific decision maker questions. While Fit for Purpose views can be used to augment any architecture viewpoint, the Human Viewpoint methodology has formalized their inclusion with the development of the Human Views (Handley 2012).

Fit for Purpose views emphasize the relationships between the individual views that were captured in the data map at the context stage of the Human Viewpoint methodology. Although the data collection stage focused on the individual view categories, the resulting Fit for Purpose models provide integrated human-focused data to capture the interactions of the socio-technical system. The data and models from the Fit for Purpose development can be continually refined and improved as the architecture effort matures. The initial iteration of the models can be used to represent the current "as-is" architecture; this baseline can then be used to evaluate proposed changes captured in subsequent iterations that represent alternative or "to-be" architectures.

Fit for Purpose presentations allows the creation of new or custom views that easily present pertinent information in a specific stakeholder context. This allows a flexible presentation of architecture data in a manner that is more meaningful to stakeholders and useful for decision making. The Fit for Purpose Human View

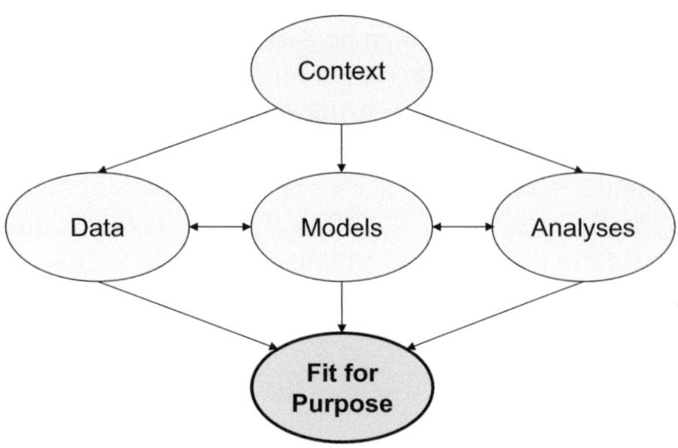

Fig. 10.1 Human Viewpoint methodology Stage 5: Fit for Purpose

models help decision makers understand the human components of a socio-technical system and provide decision data on the impact of changes to the socio-technical system on overall system performance.

10.3 Human Views—Constraints

The last Human View to be created in the sequence of view development is the Constraints view, as shown in Fig. 10.2. Constraints capture the limitations that impact the assignment and performance of personnel completing different tasks. The original Human View included six sub-views for the Constraints model; these were different sets of specific types of constraint data (Handley and Smillie 2008). Over time, this view has become much less prescriptive and much more flexible in capturing only the constraint data that is pertinent to the stakeholder concerns. Currently, the constraint data is generally broken into two subsets—constraints that impact the definition of task requirements or performance, and constraints that limit the role availability or assignment.

The original Constraints sub views are shown in Table 10.1 (Handley and Smillie 2008). At that time the Constraints view was intended to be a structured linkage from the engineering community to the Human System Integration (HSI) community. HSI considers the impact of system design on users during the system development and design process. The set of Human Views consider the impact of some HSI concerns when evaluating alternative socio-technical systems, especially from the HSI domains of Manpower, Personnel, and Training as they relate directly to the Tasks, Roles and Training Human Views.

As the role of the Human Views evolved to focus on specific areas of concern, the Constraints view was redesigned to specify sets of limitations that bounded the possible solutions presented in the Fit for Purpose views. The Career Progression and Personnel Policy sub views were removed, as these were not a good fit for the evolving role of the Human Views as an analysis tool. Both the Health Hazards and Human Characteristics sub views, which link directly to specific HSI domains, are evaluated later in the system development—these were also removed, however aspects of these can be included within the Constraints view as needed. Aspects of the Manpower Progression and Establishment Inventory are generally still captured within the Constraints view, as they apply to the specified area of concern.

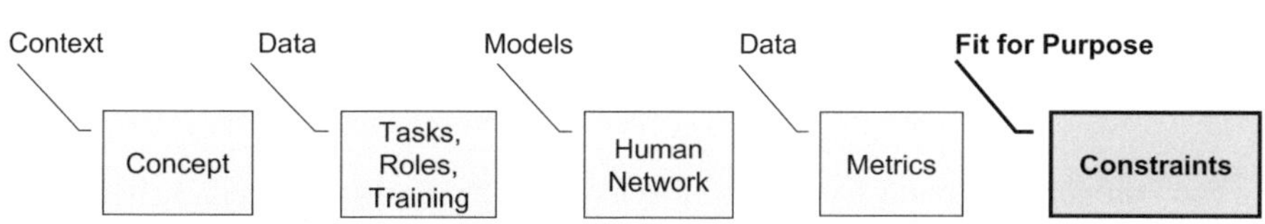

Fig. 10.2 The constraints view in the Human Views sequence

Table 10.1 The original Human View constraints sub views

Constraints sub views	Description
Manpower projections	Predicts manpower requirements for supporting present and future projects that contribute to larger capabilities
Career progression	Illustrates career progression including the essential tasks, skills, and knowledge (and proficiency level) required for a given job
Establishment inventory	Defines current number of personnel by rank and job within each establishment
Personnel policy	Defines the various department policies dealing with (governing) Human Resource issues
Health hazards	Considers the design features and operating characteristics of a system that can create significant risks of illness, injury or death
Human characteristics	Considers the physical characteristics of an operator, and movement capabilities and limitations of that operator under various conditions

The revised Constraints view includes parameters that are used to adjust the expected roles and tasks based on the capabilities and limitations of the humans in the system. While captured in a single view, constraints are generally considered either human resource constraints, that impact the roles, or human factors constraints that impact the tasks. For example, role assignments may be impacted by personnel availability while work cycles or equipment downtime may impact the frequency or assignment of tasks. Constraints bound the abilities of the human component in relation to operational demands and system components.

10.4 Stakeholder Trade-Offs

Stakeholders make trade-off decisions between cost, schedule and risk in order to achieve specific program goals. These trade-off decisions during the system architecting phase must include consideration of the human component in order to accurately assess system performance and to identify options for reducing life-cycle costs. The resources needed to train and retain the types and numbers of people who will interact with a system determines a significant portion of the long-term costs and affordability of the system. About one half of the life-cycle cost of a system is personnel related, and that cost may rise with flawed decisions concerning the selection of roles and assignment of tasks. Personnel costs are often cost multipliers for fielded systems, since multiple people are often involved with a single system as they interact with the system over its lifetime (Warner 2016).

The Human Viewpoint can be used to evaluate manpower costs and risks from a strategic approach, i.e., manpower tradeoffs can be assessed in order to address alternative personnel mixes that remain within budgetary constraints. Risk can be minimized by the availability of certified personnel at the needed time at an acceptable training cost. For example, providing qualified personnel minimizes risks due to errors or delays, but may incur salary and training costs and delay the schedule due to hiring or training time. The Fit for Purpose Human Views provide the structured data to support the cost and risk analysis of the impact on the number and type of personnel required, the need for new skills, knowledge, or competences, and changes to existing tasks and work processes.

The Human Viewpoint Fit for Purpose Views can answer specific questions for stakeholders on the human component of the system, such as: What is the impact of lower/higher skill levels than required by the task? What is the impact of role compatibility based on similar/dissimilar critical task training? What is the impact of increasing role workload due to multiple tasks? It is critical that these human system trade-offs become more rigorous in order to fully identify and mitigate the overall system risk and long-term cost impacts. The Fit for Purpose Human Views can provide stakeholders the necessary information in an easily understood format so that quantitative data can be used to objectively evaluate the alternatives, rather than choose the more expedient solution.

10.5 Example

The last stage in the Human Viewpoint methodology is to render Fit for Purpose models that allow stakeholders to easily access specific decision focused information. As part of the Fit for Purpose model development, data on personnel and task limitations is captured in the Constraints view; this data is used to set realistic expectations for the analysis of alternatives. For the communication system example developed over the last few chapters, the Fit for Purpose models can describe the different attributes of the candidate operators, as well as the predicted performance differences while operating the equipment. Figure 10.3 shows a SysML Block diagram that depicts the characteristics of the vehicle crew. Note that the role description for the third seat, the communication equipment operator, describes the two candidates and indicates the training each has received, as this is one of the main differentiators between the two role choices. Figure 10.4 shows a SysML Sequence diagram of the task "Make Softphone Calls" with the predicted timeliness performance metric for each candidate role completing the task; differences in completion time of the communication task by an Infantryman were predicted to only take 10% longer to complete than a Signal Specialist (Sargent and Walters 2015).

Additional information to inform the stakeholder decision on the appropriate candidate for the vehicle third seat is captured in the Constraints view, which provides information on the numbers and types of soldiers available, as shown in Table 10.2. The constraints data indicates that the number of personnel with the Signal Specialty is severely limited: only 4% of the soldiers available are designated as Signal Specialists while 50% of the soldiers are designated as Infantryman. There are simply not enough Signal Soldiers available to staff all the vehicles that use the new communication equipment with a Signal Soldier (Handley et al. 2015).

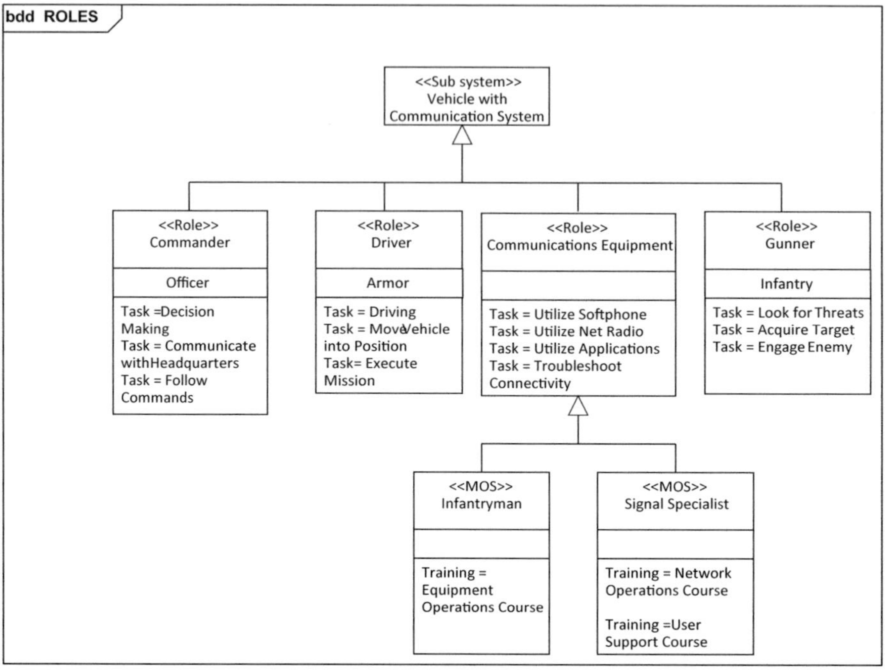

Fig. 10.3 Fit for Purpose roles view with role alternatives

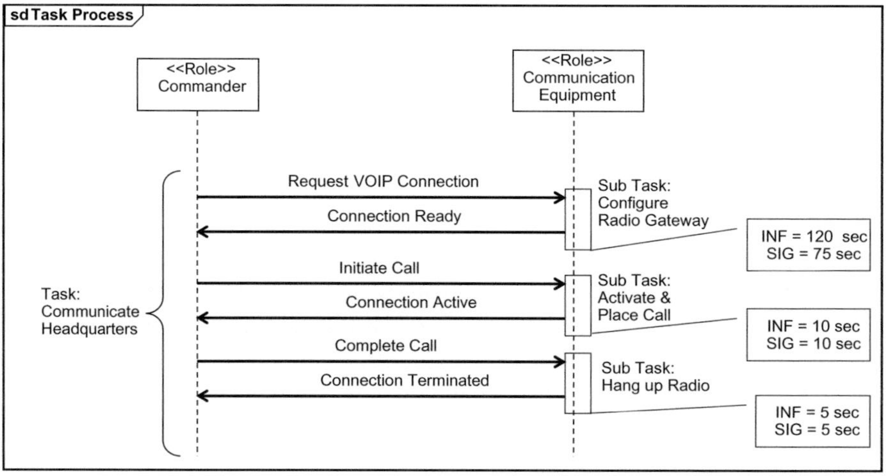

Fig. 10.4 Fit for Purpose tasks view with timeliness performance values

Table 10.2 Constraints view data for the communication system example

	Infantryman	Signal specialist	Other soldiers	Total personnel
Assigned to headquarters	12	4	36	52
Assigned to vehicles	24	0	24	48
Totals	36	4	60	100

Lastly, the analyses provided in the previous chapter suggested that increasing equipment specific training for the Infantryman and providing troubleshooting assistance through the Human Network can relieve the need for the limited availability Signal Specialist to be assigned to the equipment. Additional analyses that include human factors constraints can evaluate whether adding the additional tasking to the current crew member will overload the operator, causing a performance decrement. All of this information can be provided to system stakeholders to consider as part of the decision to determine the appropriate candidate to assign to the third seat of the vehicle crew.

10.6 Summary

The Human Viewpoint methodology results in a complete set of Human View Fit for Purpose models with context specific data and supporting analyses that can be used to evaluate different alternatives to address stakeholder concerns. Additionally, the Constraints view specifies sets of limitations that bound the possible solutions presented in the Fit for Purpose models; it includes parameters that are used to adjust the expected roles and tasks based on the capabilities and limitations of the humans in the system. The Human Viewpoint provides a repository of human focused data to support socio-technical system analyses and completes the architecture description to share with system stakeholders.

The Human Viewpoint completed for the communication equipment example provided information on alternative role assignments for the third seat in the vehicle. The analyses linked the differences in training for the communication tasks to performance impacts. This was only one set of data for stakeholders to consider; other factors, such as personnel availability and outside assistance, were also identified in the Human Views and can be considered by stakeholders for decision making. The complete set of data captured in the Human Views for the communication equipment example is shown in Table 10.3.

Table 10.3 Communication system example Human Views data

Views	Content
Concept	Vehicle crew positions
Tasks	Communication equipment tasks
Roles	Infantryman and signal specialist descriptions
Training	Basic and equipment specific training
Human network	Interactions of the communication operator with other roles
Metrics	Softphone human and system performance measures
Constraints	Infantryman and signal specialist manpower numbers

References

DoDAF (2010) The DoDAF architecture framework version 2.0. U.S. DoD Deputy Chief Information Officer. http://dodcio.defense.gov/Portals/0/Documents/DODAF/DoDAF_v2-02_web.pdf. Accessed 1 Nov 2018

Handley H (2012) Incorporating the NATO human view in the DoDAF 2.0 meta model. Syst Eng 15(1):108–117

Handley H, Smillie R (2008) Architecture framework human view: the NATO approach. Syst Eng 11(2):156–164

Handley H, Kandemir C, Dario E (2015) Warrior information network-tactical human view models & analysis. Technical report, Old Dominion University, Norfolk, VA

Sargent B, Walters R (2015) Human performance model results. Technical report, Alion Science & Technology, Boulder, CO

Warner J (2016). The cost of not accommodating the warfighter. Technical report, Headquarters Department of the Army, Washington, DC

Chapter 11
Simulation Models—The Human View Dynamics

Abstract This chapter discusses the use of simulation models and the original Human View Dynamics. Simulation models can be used to provide time-based analyses of the socio-technical system and evaluate its response to different external stimuli. The original set of Human Views included a dynamic simulation model to augment the static set of views. While this view is now optional, it is encouraged to perform simulations with a tool such as IMPRINT to capture socio-technical system behavior in order to evaluate alternative system configurations based on performance metrics.

Keywords Simulation model · Human View Dynamics · IMPRINT

11.1 Introduction

Along with the seven Human View models, the original Human Viewpoint description recommended an accompanying simulation model termed the Human View Dynamics. The purpose of the dynamic view was to captures aspects of the socio-technical system components defined in the other Human Views and simulate the resulting composite model in response to external stimuli to evaluate the system performance over time. The results can be used to inform stakeholder decisions and included in the Fit for Purpose views. The Improved Performance Research Integration Tool (IMPRINT) been used as a simulation environment for creating a human performance model based on the Human Views data. While not a requirement of the Human Viewpoint methodology, completing the optional Human View Dynamics provides the opportunity to vary different sets of data over multiple simulations to evaluate the performance of the socio-technical system.

H. A. H. Handley, *The Human Viewpoint for System Architectures*, Topics in Safety, Risk, Reliability and Quality 35, https://doi.org/10.1007/978-3-030-11629-3_11

11.2 The Human View Dynamics

The original Human Viewpoint definition included an eighth product, the Human Dynamics. The intention of this dynamic Human View was to capture the interaction of the socio-technical system components defined in the other products (Handley and Smillie 2008). While this model could be a static view that depicted the behavior of the human system, for example a state machine, it was recommended to be a simulation model that could provide performance data. However, a simulation model did not fit with the portfolio of existing system architecture views, therefore the Human Dynamics, as part of the Human Viewpoint, was removed. However, human performance simulations are still an important part of the architecting process, and a requirement for most Fit for Purpose views in order to provide stakeholders human performance data.

The objective of the Human View Dynamics is to create a simulation model that can capture socio-technical system behavior in such a way that it can be used to evaluate alternatives based on performance metrics. A simulation model can predict the impact of different operator attributes on system performance, as well as impacts from the system demands on the operator. A schema of the interaction of the data captured in the Human Views that can be realized in a simulation model is shown in Fig. 11.1. Each Human View model captures a different set of human elements; the simulation model triggers the relationships between the elements to predict system performance.

As shown in Fig. 11.1, an event from the concept scenario triggers a task. The role responsible for the task begins processing it. The role may coordinate with other crew members via information exchanges while processing the task. The ability of the role to complete the task with acceptable performance may depend on training and impacted by certain constraints. Use of a system resource, or interface, may be required to complete the task and is included in the model. Once the task

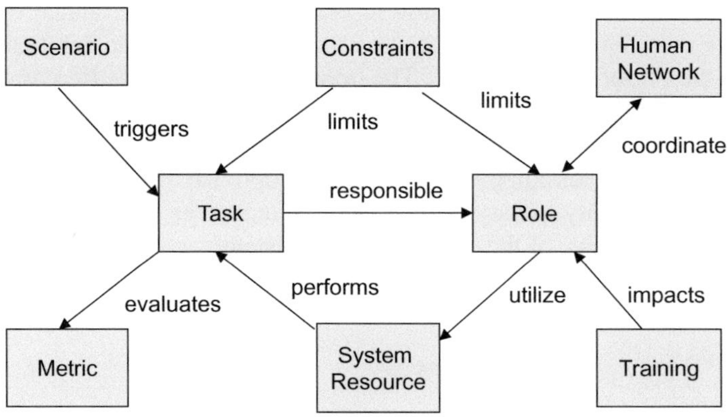

Fig. 11.1 The Human View Dynamics schema

is completed; metrics are used to evaluate the task outcomes. This schema of inter-view relationships can be used to develop a simulation model to evaluate the impact of different socio-technical system alternatives on the system performance. The simulation outcomes can then be used to augment Fit for Purpose views and inform decision maker trade off analyses.

11.3 IMPRINT

The Improved Performance Research Integration Tool (IMPRINT) is a dynamic, stochastic discrete event network modeling tool designed to help evaluate socio-technical systems (Mitchell and Brennan 2008). IMPRINT incorporates task analysis, workload modeling, environmental stressors, and embedded personnel characteristics to perform simulations of humans interacting with technology. Data are entered through user interfaces and task network diagrams; underlying human performance algorithms are then employed to perform simulations. Performance time and accuracy requirements are collected and workload profiles are generated so that role to workload distribution and role to system task allocation can be examined (IMPRINT 2007).

The data captured in the Humans Views can be aligned with the inputs required by the IMPRINT model, as shown in Table 11.1 (Handley and Broznak 2011). Configuring IMPRINT requires identifying the roles, mapping them to tasks, and indicating the system interface requirements. The simulation outcomes then capture the impact of the tasking on the role, through workload metrics, and the role's impact on the system, through task performance measures. By configuring an IMPRINT simulation model with the Human Views data, the information captured in the Human Viewpoint can be used to support a dynamic evaluation of the socio-technical system performance.

An important feature of IMPRINT is its ability to evaluate mental workload while simulating operators interacting in a task process. This capability is important because the amount of mental workload that is required has a significant effect on human performance. Wickens' Multiple Resource Theory is the basis for the IMPRINT workload algorithm (Wickens 1991). When performing multiple tasks at the same time, the operator is utilizing the same limited resources for the concurrent tasks; this combination of limited cognitive resources and multiple task demands may result in high workload that leads to a greater number of errors, increased task time, or both. Simulations allow input parameters to be varied, constraints to be relaxed and other variables affecting human performance to be explored in order to evaluate alternative socio-technical system designs.

Table 11.1 Mapping of the Human Views data to IMPRINT requirements

Human Views data		IMPRINT requirements
Concept	High level scenarios or use-cases	Triggers to simulate the model
Tasks	Task decomposition and interdependencies; systems available for task completion	Network diagram composed of tasks and subtasks; assignment of system interfaces to tasks
Roles	Role definitions and assigned task	List of available operators; assignment of operators to tasks
Training	Training types required to perform assigned tasks	Training moderator settings
Human network	Interactions required between roles to complete tasks	Identification of communication and coordination functions
Metrics	Performance parameters and standards	Mission level time and accuracy criteria and task level time and accuracy standards
Constraints	Role and task limitations under various conditions	Personnel moderator settings and stressors

11.4 Example

IMPRINT was used to develop a Human Dynamics model to support the Human Viewpoint development for an operations center that included virtual teams (Handley et al. 2006). Virtual teams exist when decision-making activities are distributed across a team and the team is also distributed across physical locations. Reach-back occurs when an external role "reaches back" for information or services. Human centered aspects of reach-back, such as differences in operational tempo and priorities, can affect the performance of a task process performed by a virtual team. A dynamic model including the socio-technical data can be used to assess the impact of the virtual team on the work process performance.

The Human Viewpoint development collected information on the tasks, roles and work processes for the operations center. This provided the data used to populate the IMPRINT model. The IMPRINT model represents the operations center staff tracking current activities as well as planning future activities, plus monitoring communications and updating their shared situational awareness. The model includes the impact of having collaborating personnel not co-located with the rest of the team on task processes. A series of simulations were conducted to vary the reach-back status of different roles in the organization; the results were used to evaluate the impact of reach-back on team outcomes.

The IMPRINT model focused on the "Course of Action Planning" process, which can be decomposed into three top level functions: "Mission Analysis", "Course of Action Development", and "Provide Plans and Orders". An example of the task process and the interaction of the roles, is shown in the sequence diagram of Fig. 11.2.

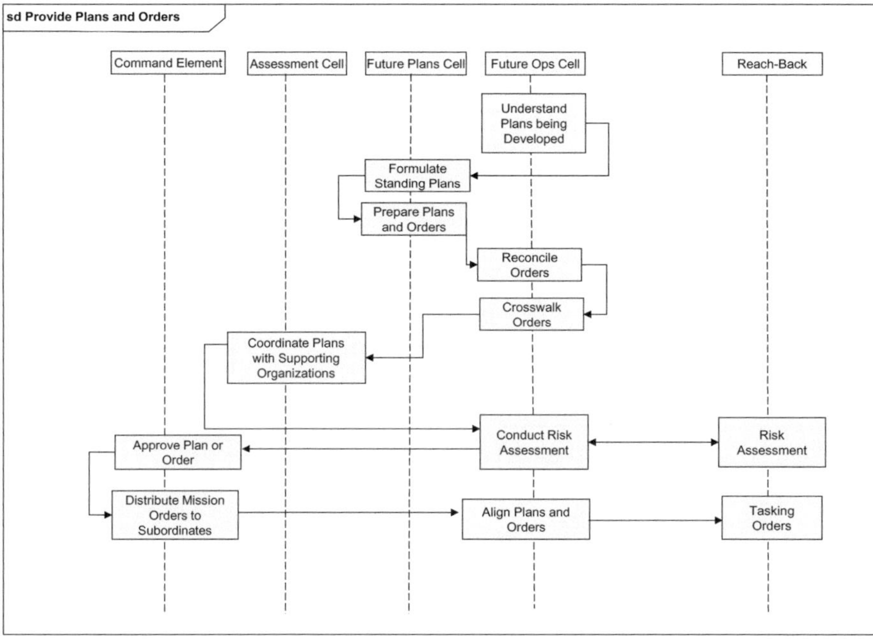

Fig. 11.2 Human View tasks sequence diagram for "Provide Plans and Orders"

The diagram captures the third function, "Provide Plans and Orders" and indicates the sequence of tasks and the responsible roles.

The top-level IMPRINT model for the operations center is shown in Fig. 11.3. The IMPRINT model captures the sequence of the three high-level functions and the decomposition of each function into the underlying task processes. Figure 11.4 depicts the detailed IMPRINT tasks for the "Provide Plans and Orders" function. Each of the tasks captured in the model is further customized with additional data, such as the expected delay time, workload values, the assigned operator, and other parameters as shown in Table 11.2.

The workload experienced by the operator in this model comes from three distinct sources. The first is a baseline workload from the communication monitoring function. This is constant throughout the scenario as the operator checks for communications. When a communication is received by an operator, this adds workload to the baseline value depending on the type of communication and the type of response required. Direct communications are triggered by the task process during or at the completion of tasks; the direct communications experience additional workload in a reach-back situation due to the increased cognitive load of the operator. The third workload component is from the work process task that the operator completes.

The IMPRINT model can be configured to use the correct parameters when the interactions between team members are either in a co-located or a reach-back condition. Cognitive load is increased when the team members are more interdependent

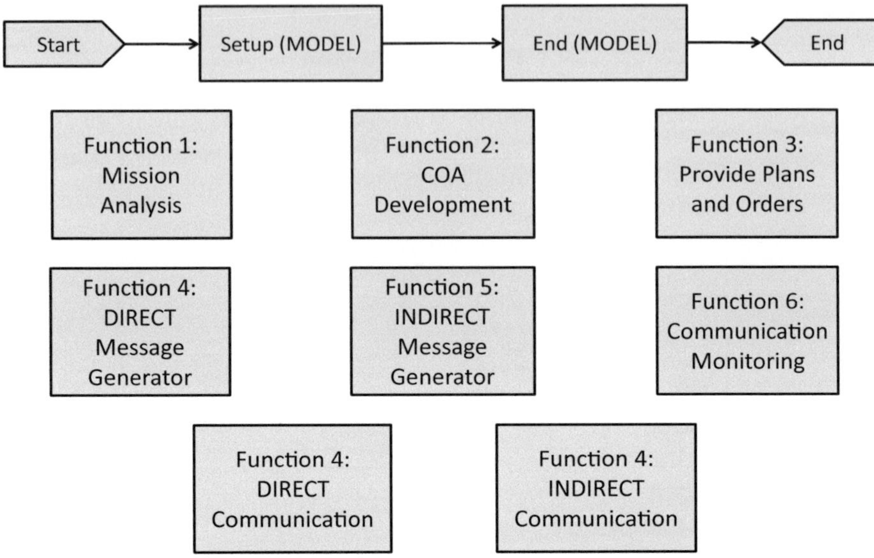

Fig. 11.3 Top level IMPRINT model functions for course of action planning example

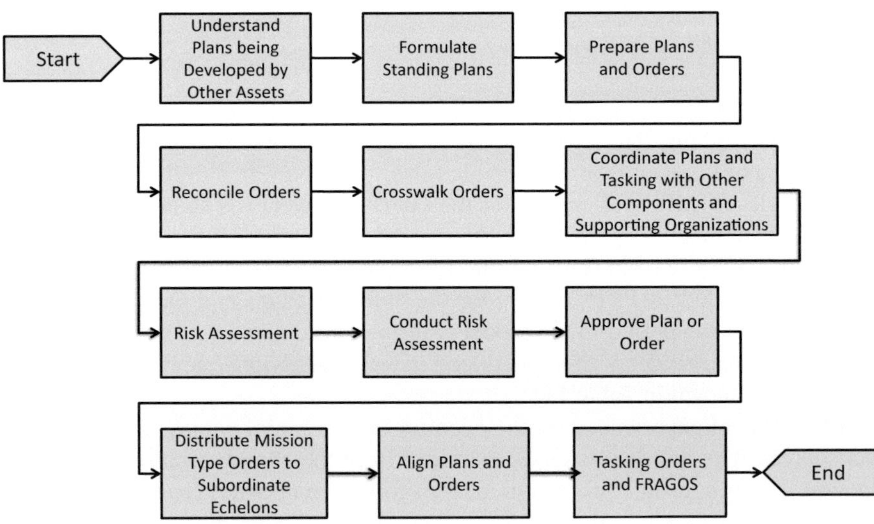

Fig. 11.4 Provide Plans and Orders IMPRINT model tasks

Table 11.2 Sample IMPRINT task configuration data

3 Provide Plans and Orders	Operator assigned	Communications required	Timeliness value
Understand plans being developed by other assets	Future Ops	Direction	Random (5,15)
Formulate standing plans	Future plans	Coordination	Random (60,300)
Prepare plans and orders	Future plans	Information	Random (300,600)
Reconcile orders	Future Ops	Coordination	Random (60,600)
Crosswalk orders	Future Ops	Coordination	Random (5,15)
Coordinate plans and tasking w/other components	Assessment	Coordination	Random (105,315)
Risk assessment	Reach-back	Information	Random (5,15)
Conduct risk assessment	Future Ops	Coordination	Random (105,315)
Approve plan or order	Command	Information	Random (5,15)
Distribute mission type orders to subordinate echelons	Command	Information	Random (5,15)
Align plans and orders	Future Ops	Direction	Random (5,15)
Tasking orders	Reach-back	Direction	Random (105,315)

Table 11.3 IMPRINT model reach-back experimental conditions

	Increasing levels of reach-back	
	Co-locate	Reach-back
Control	CE, FOC, AC, FPC, RC	None
Trial 1	CE, FOC, AC, FPC,	RC
Trial 2	CE, FOC, AC,	RC, FPC,
Trial 3	CE, FOC,	RC, FPC, AC,

CE Command Element, *FO* Future Ops Cell, *AC* Assessment Cell, *FPC* Future Plans Cell, *RC* Remote Cell

and therefore require more interaction; this is more pronounced in a reach-back situation. The change in operator workload impacts the timeliness and accuracy of the task process being simulated. Table 11.3 shows the experimental conditions that were used to execute the IMPRINT simulations. The locations of the team members were varied over different simulations to assess the impact of reach-back on the socio-technical system.

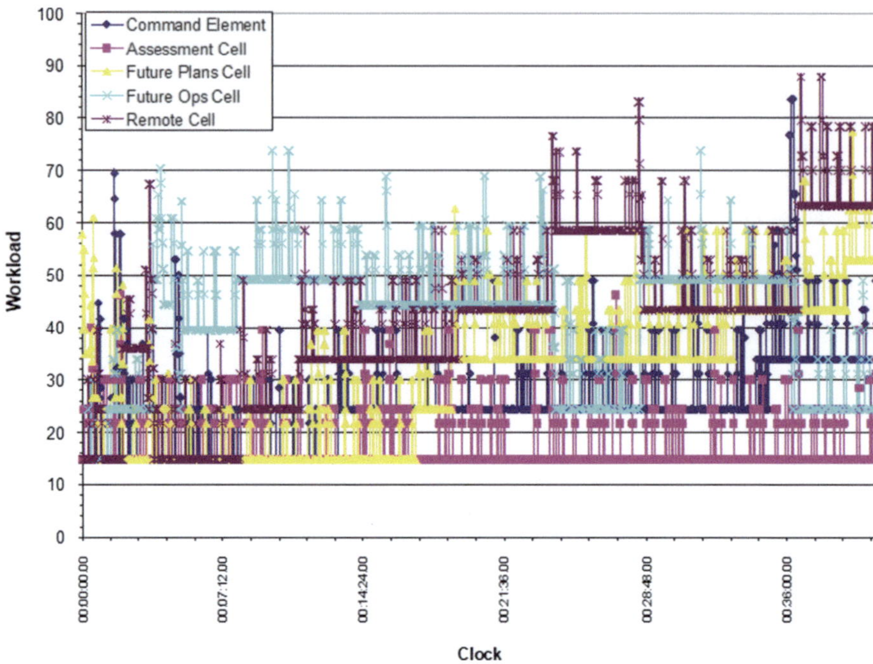

Fig. 11.5 Example IMPRINT simulation results of workload over time

An example of an IMPRINT output graph that depicts the changes in workload over time for each role is shown in Fig. 11.5. The simulation results indicate that when a role is in a reach-back position the workload of the other roles communicating with this role increase due to the need for direct communications. However, in most cases the increase is small and does not significantly raise the overall workload. The exception is for roles that are communicating frequently over small periods of time or are already highly tasked. In these cases, the additional workload is enough to send the roles into an overloaded condition. This is further exacerbated in stressed conditions, i.e. under an increased operational tempo. In this case the communication patterns and tasking may need to be adjusted to balance the workload among the roles.

11.5 Summary

The Human Views collect information pertaining to the socio-technical system; the Human Dynamics is a simulation model that can evaluate the interaction and performance of the data captured in these views. The Improved Performance Research Integration Tool (IMPRINT), a human performance modeling tool, has been used as the basis for the development of the Human Dynamics by creating an alignment between the data collected in the Human Views and the input parameters required by the IMPRINT model. IMPRINT provides a comprehensive model which easily maps to the data defined in the Human Views.

Dynamic simulations can be used to evaluate time-based system performance and the influence of human parameters on the predicted system outcomes. Typically, dynamic models allow input parameters to be varied, constraints to be relaxed and other variables affecting the performance of the human operator to be explored in order to evaluate the socio-technical system design. The example simulation model described in this chapter, the distributed operations center, illustrated how the Human Dynamics can be used to evaluate an area of stakeholder concern, such as the impact of reach-back roles collaborating in a work process. While no longer a requirement of the Human Viewpoint methodology, the use of a human performance simulation is still necessary for most Fit for Purpose view development in order to provide stakeholders relevant human performance data.

References

Handley H, Broznak D (2011) Developing, enhancing and utilizing human system integration tools—Task 7: Human view processes and measures for role analyses. Technical report, Old Dominion University, Norfolk, VA

Handley H, Smillie, R (2008) Architecture framework human view: the NATO approach. Syst Eng 11(2):156–164

Handley H, Smillie R (2010) Human view dynamics—the NATO approach. Syst Eng 13(1):72–79

Handley H, Sorber T, Dunaway D (2006) Maritime headquarters with maritime operations center, concept based assessment. Technical report, Pacific Science & Engineering Group, San Diego, CA

IMPRINT (2007) IMPRINT PRO analysis guide. Technical report, U.S. Army Research Laboratory, Human Research and Engineering Directorate, Aberdeen Proving Ground, MD

Mitchell D, Brennan G (2008) Mission centered human system analysis using FCS vignettes. Technical report, Army Research Laboratory, Human Research and Engineering Directorate, Aberdeen Proving Ground, MD

Wickens C (1991) Processing resources and attention. In: Damos D (ed) Multiple task performance. Taylor & Francis, Washington, DC, pp 3–34

Chapter 12
Applying Human Viewpoints to Risk-Based Decision-Making

Alan Bell
alan.bell@wings-of-gold.com

Abstract This chapter provides a detailed example of an application of the use of the Human Views in risk-based decision making. The use case steps through the Human Viewpoint methodology and renders SysML activity diagrams, which can be aligned with an influence diagram, to identify the uncertainties in the system. By creating models for both the As-Is and To-Be systems states, the methodology can be used to support a variety of data-driven, quantitative analysis to provide decision support for a proposed change's influence on the performance of a complex system.

Keywords Risk-based decision-making · Influence diagrams · As-Is and To-Be states

12.1 Introduction

The Federal Aviation Administration (FAA) is responsible for ensuring safe and efficient air traffic management operations throughout the US National Airspace System. During a twenty-year period beginning at the turn of the century, wind farms were constructed in many states throughout the country, with some of these farms being located in the vicinity of airports and concurrently, in the vicinity of the surveillance RADARs used by air traffic controllers to sequence and separate aircraft. This case study explores the influence of these wind farms on aviation safety and demonstrates an application of the Human Viewpoint methodology for use in quantifying risk in support of management decisions about the airspace system and its associated architecture.

12.2 Background

Over the past two decades, wind energy has grown substantially as a source of renewable power in the United States. The rate of growth between 2001 and 2018 is depicted graphically in Fig. 12.1.

H. A. H. Handley, *The Human Viewpoint for System Architectures*, Topics in Safety, Risk, Reliability and Quality 35, https://doi.org/10.1007/978-3-030-11629-3_12

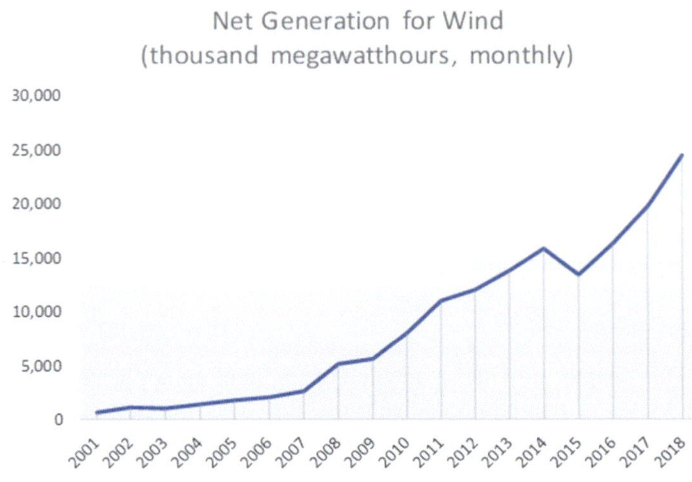

Fig. 12.1 Wind energy growth, 2001–2018

To generate this energy, wind farms consisting of many large tower-mounted wind turbines have been erected in several states, with the turbine blades extending several hundred feet in the air. In some cases, the height of the towers and the proximity of the farm to local airports has resulted in the turbine blades being visible to the FAA's surveillance RADAR. This condition can lead to a misinterpretation of the reflected energy, causing wind turbines to be mistaken for aircraft, both by air traffic controllers and the automation systems that support their activities.

12.3 Context

To provide the context for the risk-based decision faced by FAA engineering managers, high-level diagrams may be used to define the scope of the analysis and provide an overview of the system. This technique allows definition of not only the system as it exists, referred to as the As-Is state, but also any historic As-Was states that might be relevant to the analysis, as well as any future To-Be states that would result if potential changes currently under consideration are implemented.

Air traffic controllers use RADAR to determine the position of aircraft operating in the vicinity of a local airport. With each sweep of the RADAR, an automation system known as Standard Terminal Automation Replacement System (STARS) updates the position of each aircraft and presents the information to the controller via a display. This allows the controller to monitor the progress of the arrivals and project conflicts that may develop as aircraft approaching from a variety of directions merge onto concurrent routes. If a controller projects a conflict, interventions such as changes to heading, altitude, or speed can be issued to avoid a collision. The

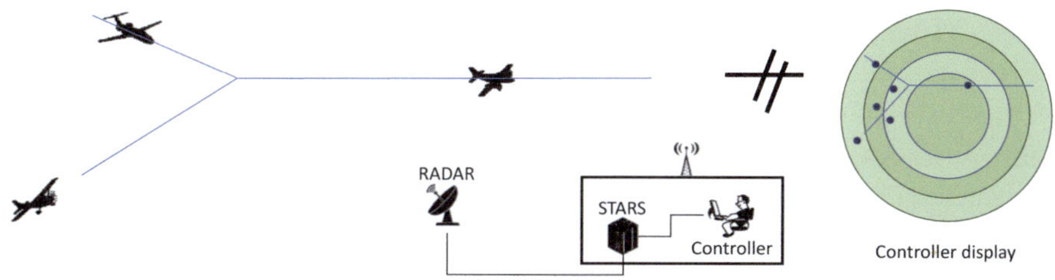

Fig. 12.2 As-Was air traffic management environment

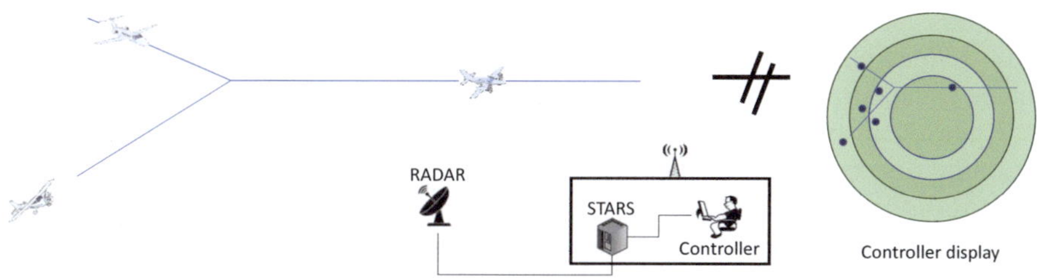

Fig. 12.3 As-Is air traffic management environment

STARS system assists the controller with conflict detection by generating an alert if its algorithms detect a conflict between any two aircraft. Figure 12.2 presents a general overview of the environment as it existed before the emergence of the wind turbines and includes a depiction of the display that might be presented to a controller under the circumstances depicted.

Figure 12.3 is an updated version of Fig. 12.2 showing the introduction of the wind turbines and the effect this change to the environment causes on the controller's display.

In Fig. 12.3, the addition of the wind turbines has created additional contacts on the controller's display as the RADAR can't distinguish between signals reflected from real aircraft versus signals reflected from the turbine blades. These false contacts also result in STARS producing a much larger volume of alerts to the controller, many of which are false alerts.

In the wake of wind farms being established near airports in several locations, the frequency of accidents in this environment has increased and air traffic controllers have reported a reduced level of confidence in the conflict alerts. FAA executives have hypothesized that adding a filter to STARS would reduce the number of false contacts, thereby reducing the number of false alerts, and restoring accident frequency to the pre-windmill rate.

12.4 Models

A system architecture is an excellent tool for developing an understanding of how a system works, when it works as it's designed to work, and for managing data associated with the system's operation and performance. However, to demonstrate or assess how a system can be degraded or fail altogether, the architecture must typically be augmented with additional models to define the scope of analysis, assist with the identification of uncertain performance that may reside within relevant portions of the system, and to capture the relationships between uncertain events that allow analysis of the area of concern. Using standardized modeling formats such as those defined within the System Modeling Language (SysML), additional diagrams can be created to unlock quantitative analytic tools.

Figure 12.4 uses a SysML activity diagram to present a depiction of the As-Is air traffic management environment shown earlier in Fig. 12.3. The graphic shows RADAR information entering the system and being processed by STARS, then sent to the controller's display. The processed track information is also evaluated by STARS, and an alert is provided to the controller if certain conditions are satisfied.

The human controller is resident within the system and once the information from STARS is received, it is interpreted, and the expected future tracks of the aircraft are used to project conflicts. According to the controller's training and motivation to prevent accidents, if the controller becomes aware of a potential conflict, an intervention will be provided to avoid a collision.

If management chooses to add a filter within STARS and activate it, this new functionality can be added to the As-Is model, creating a potential To-Be model. Doing so creates a new activity diagram such as the one shown in Fig. 12.5, with the new filter functionality shown in a lighter shade.

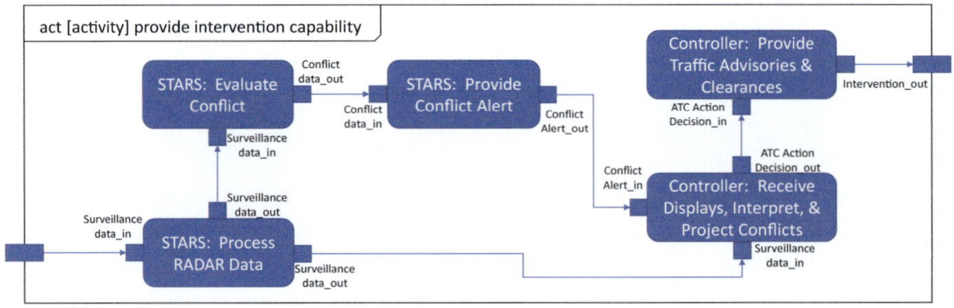

Fig. 12.4 SysML activity diagram for As Is environment

12.5 Analysis

Once the As-Is and To-Be models of the system are completed, the next step in applying the Human Viewpoint methodology is to identify uncertain performance within the system, including human performance. This is accomplished by consulting with subject matter experts to gain deeper insight into the nature of the functions inherent to the system and developing an understanding of how the system might fail to perform as intended.

For the purpose of illustration, Fig. 12.5 is updated with brief questions that help to identify uncertain performance associated with each of the system functions. The result of this exercise is shown in Fig. 12.6.

Once uncertainties have been identified, they may be represented graphically in an influence diagram. An influence diagram uses symbols to indicate decisions, uncertain events, deterministic events, and potential effects. In Fig. 12.7, the questions presented in Fig. 12.6 are replaced by ovals with key words used as labels to represent uncertainties, while the question regarding whether or not to activate the filter, which is a decision, is represented with a rectangle.

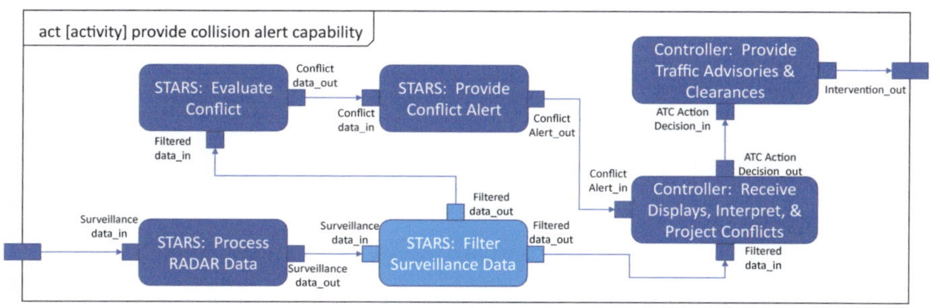

Fig. 12.5 SysML activity diagram for the To-Be environment

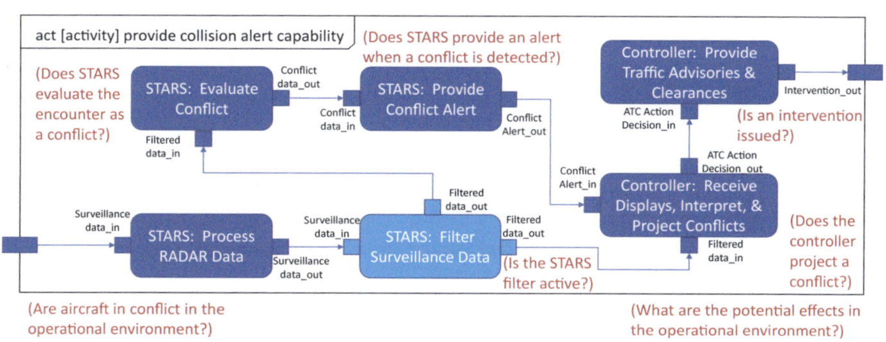

Fig. 12.6 Identification of uncertainties

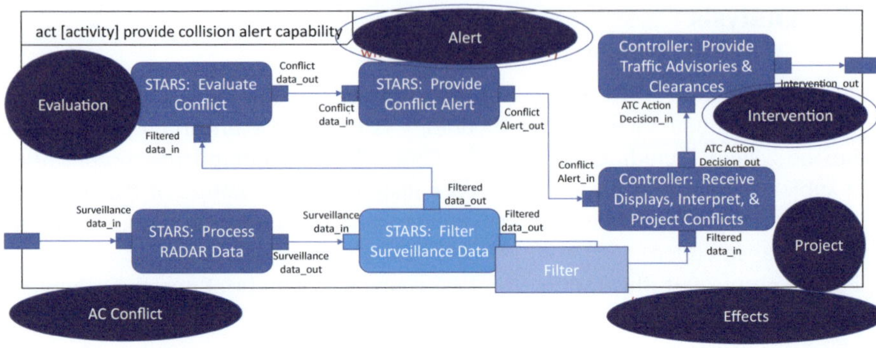

Fig. 12.7 Symbolic representation of uncertain events

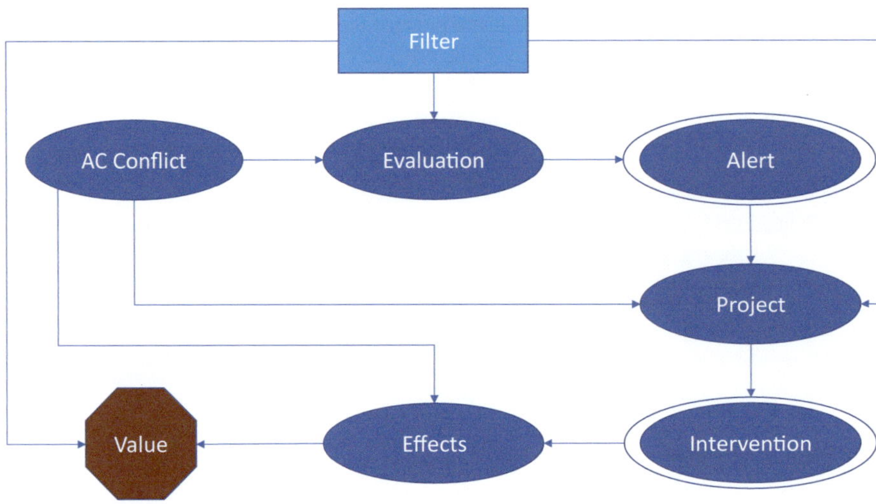

Fig. 12.8 Influence diagram for To-Be system configuration

Once the uncertainties are identified and represented symbolically, the symbols can be extracted and used to create an influence diagram such as the one shown in Fig. 12.8.

As a brief explanation of the diagram, the arrows between events represent a potential dependency that may exist. Double circles represent deterministic events, while the octagon labeled value is used to assess the costs associated with implementation and negative effects, or alternatively, the benefit associated with positive effects. The position of the symbols is meaningless, and they are arranged solely for convenience and clarity. It is also interesting to note that the As-Is system influence diagram is identical to the To-Be diagram, only it does not include the decision regarding the filter since it does not yet exist.

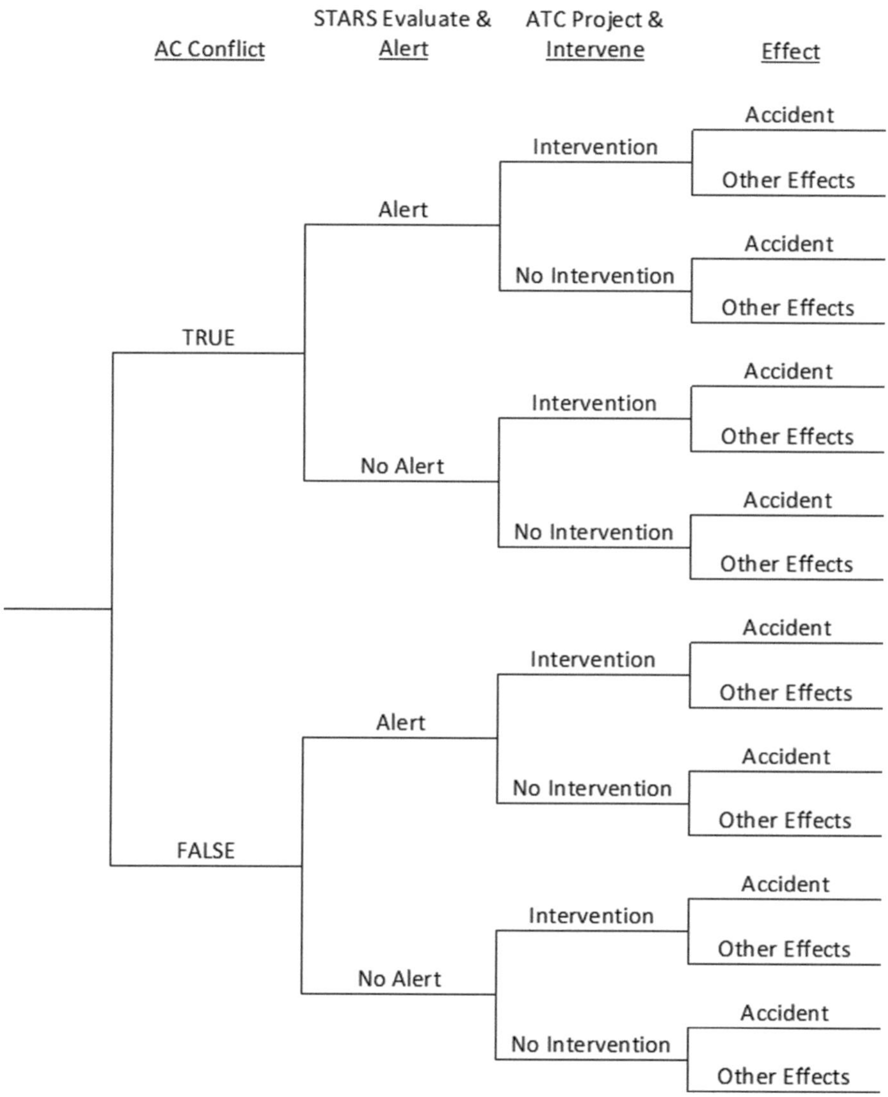

Fig. 12.9 Risk framework for system analysis

A risk framework is a specialized event tree used for quantifying risk. (In the field of decision analysis, a risk framework is a decision tree associated with a single alternative). Once an influence diagram is developed, it can be used as a blueprint for a risk framework that accounts for any of the dependencies that may exist. Figure 12.9 illustrates the risk framework that can be created from the influence diagram shown in Fig. 12.8.

In this example, the FAA is the primary stakeholder, and the effect of interest is an accident. Other effects may include near mid-air collisions, separation losses, degraded interventions, or other similar lower severity events in which no injuries or damage occurs.

Once the risk framework is complete, each branch of the event tree must be populated with data, and this requirement effectively creates a *data shopping list*. Identifying the data needed to populate the risk framework completes Part 1 of the analytic process and allows the Human Viewpoint methodology to progress.

12.6 Data

At first glance, the process of collecting data may seem overwhelming due to the volume of data necessary to complete the analysis. However, the characteristics of a risk framework include conditions in which all events in the tree are both collectively exhaustive and mutually exclusive. This means that if the probability of one binary branch of a tree is known, the other branch is automatically known due to the rule of complements. In Fig. 12.9, that immediately reduces the number of values needed to populate the data fields from 30 to 15. Similarly, analysts can take advantage of dependencies to calculate probabilities using rationalist arguments. For instance, if a real aircraft is in conflict with another real aircraft (a *true* conflict), then there is some probability of collision. However, if the aircraft is not in conflict with another real aircraft, but instead with a false target (a false conflict), then there is no possibility of a collision. This allows a rationale conclusion that the conditional probability of an accident is zero in such a circumstance, and zero may be assigned to each field associated with an accident possibility in the lower half of the framework.

Another advantage of using a risk framework to identify data needs is the focus it provides to the data collection effort. Without a risk framework, relevant pieces of data might not be obtained or could be overlooked even if available, or in the other extreme, irrelevant data may be introduced. The framework also allows insight into the best sources of data, whether from empirical observations, rationalistic calculations, or subjective estimates by experts. For instance, determining the likelihood of two aircraft being in conflict as they approach the airport could be determined using empirical methods, counting the number of aircraft pairs that arrive, and then comparing that value to the number of conflicts. Alternatively, an analytic method could be used in which the conflict rate is varied from zero to any desired value up to 1.0, allowing results to be expressed in the form of a curve rather than a point estimate.

For the purpose of this analysis, data was obtained from FAA databases that track the number and type of operations within the airspace system, RADAR performance data provided by technical experts, air traffic control subject matter experts, and safety data provided by the National Transportation Safety Board. The FAA data set consisted of 1.3 billion operations conducted over a five-year period. Additionally, data was obtained for three system configurations:

Table 12.1 Data summary

Data item	As-Was	As-Is	To-Be
P(conflict)	0.1	0.1	0.1
P(alert | conflict)	0.9933	0.9933	0.989
P(alert | no conflict)	0.0002	0.0006	0.00027
P(intervention | conflict, alert)	0.9973	0.996	0.9967
P(intervention | conflict, no alert)	0.6	0.6	0.6
P(intervention | no conflict, alert)	0.68	0.66	0.67
P(intervention | no conflict, no alert)	0.001	0.001	0.001
P(accident | conflict, intervention)	7.77E−09	7.77E−09	7.77E−09
P(accident | conflict, no intervention)	5.74E−06	5.74E−06	5.74E−06

- As-Was system: prior to introduction of wind turbines
- As-Is system: after introduction of wind turbines but prior to any risk mitigation
- To-Be system: after implementation of RADAR filters to mitigate false alert rates

A summary of the relevant data for each configuration is provided in Table 12.1. The next step in applying the Human Viewpoint methodology is to return to the analysis and populate the risk framework with data to allow risk calculations.

12.7 Analysis, Revisited

Once the items on the data shopping list have been acquired, they can be used to populate the data fields in the risk framework, allowing the necessary risk calculations to be completed. In general, the risk framework is designed to provide three different types of probabilities that may be useful to decision-makers. Many of the data items in Table 12.1 are *conditional* probabilities, and once completed, the risk framework will include several others. The risk framework also allows calculation of the *joint* probability of every possible pathway through the tree, and by summing together each of the *joint* probabilities where the outcome of the pathway is an accident, the *total* probability of an accident can be determined.

Figure 12.10 provides the complete risk framework for the historic, or As-Was environment. By summing the joint probabilities of an accident together, the total probability of an accident can be calculated and compared to historic rates for the same period the data was collected from to determine whether the analytic results are a reasonable match with the historic system performance. This is an important step

Fig. 12.10 As-Was risk framework

in the analytic process because if the models produce a result that is well beyond the boundary of the desired error margin, it may be an indication that a relevant uncertain event was not captured adequately in the model. In contrast, if the historic models do achieve an acceptable measure of validation, decision-makers will be able to more easily justify their confidence in any decision supported by the analysis. The importance of this step cannot be overstated, especially in cases where the analytic results contradict the intuitive expectations of the manager who commissioned the study.

Fig. 12.11 As-Is risk framework

The next step in the analysis is to populate the risk framework with data from both the As-Is and the potential To-Be environments. Those completed risk frameworks are presented in Fig. 12.11 and Fig. 12.12, respectively.

With all three risk frameworks complete, a wealth of information can be extracted through a variety of calculations or observations. The next step in the Human Viewpoint methodology is to determine the management objective, and state results that will be material to the decisions being considered. In this case study, the decision on the table is whether FAA executives should invest in a new RADAR filtering function.

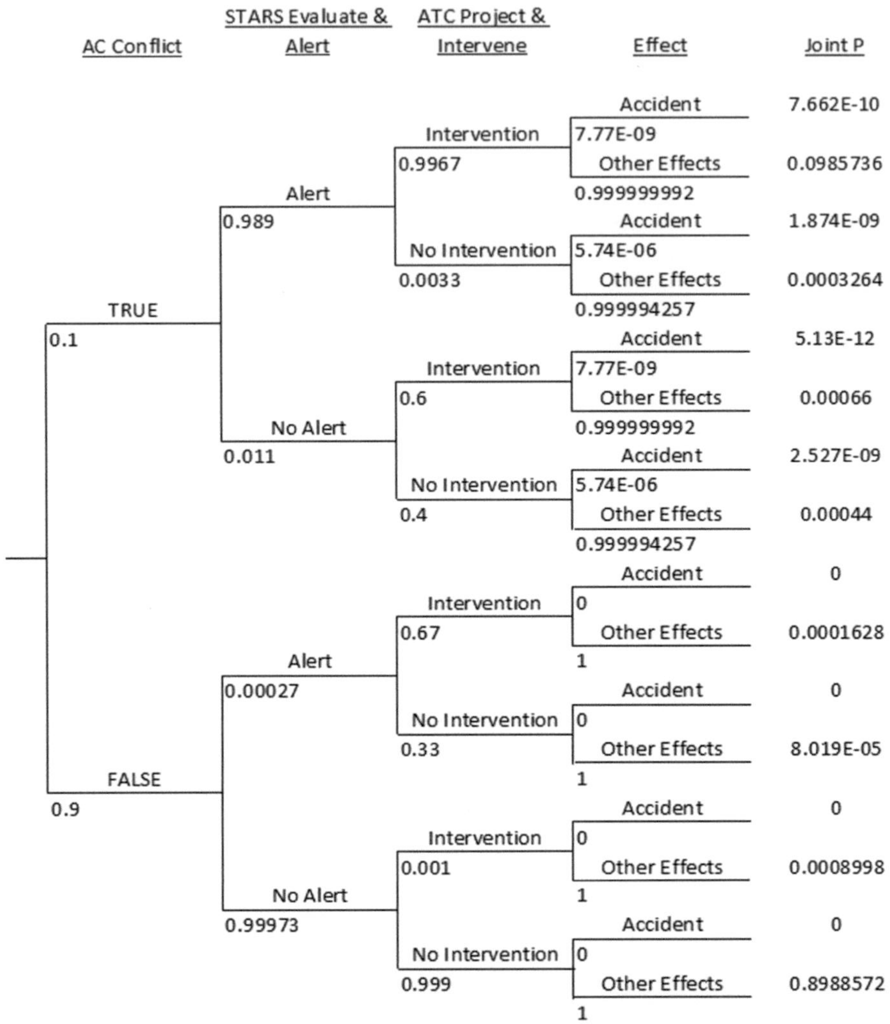

Fig. 12.12 To-Be risk framework

12.8 Fit for Purpose

Regardless of the quality of the information produced through the previous steps, it is of little value if the products cannot be used for their intended purpose. Thus, a core process within the Human Viewpoint methodology is to evaluate whether the data, models, and analysis are fit for their purpose. This includes ensuring that the initiating research question or problem can be identified within the model, and that results are relevant to management decisions. For this case study, FAA executives are interested in using the Human Viewpoint methodology to inform their investment decision regarding the RADAR filter.

Prior to discussing any decision about the addition of new functionality within the architecture to mitigate aviation safety risk, a precursor step is to ensure the problem itself can be identified within the model. By comparing Figs. 12.10 and 12.11, and evaluating the conditional probabilities within the respective frameworks, it can be seen that the probability of a false alert, an event that occurs when a false conflict leads to an alert, has risen to 0.0006. The increase in this value indicates the number of false alerts is now roughly three times higher than in the As-Was environment. Another important observation is the conditional probability of an intervention in cases when aircraft are involved in a true conflict and an alert is generated by STARS. This value has declined to 0.996, indicating that controller confidence in the alerts has diminished, and they are slightly less likely to issue an intervention in response to a true alert in the current environment. Together, these values provide quantification of two problems created by the emerging wind farms.

Finally, by summing the four non-zero joint probabilities of an accident in each framework, the result shows an increase in the risk or an accident from 3.85 e−9 in the As-Was environment to 4.59 e−9 in the As-Is environment. Stated another way, the accident rate per 1.3 billion operations (the number of operations in the FAA data set) increased from 5 to 6.

12.9 Discussion

One objective of the proposed change is to reduce the false alert rate due to false contacts created by RADAR returns from the wind turbines. Comparing the conditional probability of an alert given a false conflict in the As-Is environment to that of the To-Be shows a substantial reduction in this probability, decreasing from 0.0006 to 0.00027, a value that is nearly as low as the original condition in the As-Was framework. An additional objective is to restore controller confidence in the alerts that are generated by STARS. Once again, a comparison between the As-Is and the To-Be risk frameworks shows that this objective is largely achieved by the proposed change, with the conditional probability of an intervention given a true conflict and an alert increasing from 0.9960 to 0.9967. This change in value improves the true intervention rate, although it doesn't quite return it to its original value of 0.9973. Finally, the probability of an accident given a true alert is reduced as the system moves from the As-Is to the To-Be state. Taken together, this collection of results aligns with the FAA's qualitative assessment and provides a quantitative validation of their original findings.

These results also present and opportunity to demonstrate an advantage of the Human Viewpoint methodology over legacy methods. That is, it allows a more complete set of findings, including results that may be unexpected. Intuition dictates that if an increased number of false alerts reduced controller confidence and led to an increased accident rate, a reduction in the number of false alerts that significantly

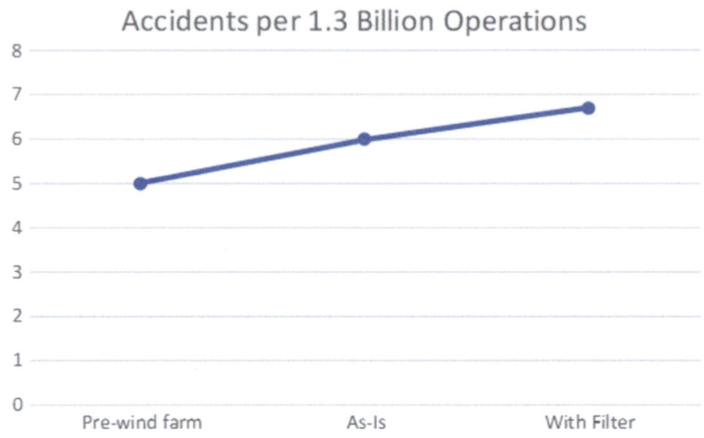

Fig. 12.13 Analysis results

restores controller confidence should reduce the accident rate to a value approaching the original rate. However, the analytic results contradict intuition, and upon close examination, the reason for the counter-intuitive result becomes clear.

As with the previous comparison, summing all joint probabilities of an accident in the To-Be environment leads to a calculation of the total probability of an accident, and in this case, a result of 5.17 e−9. This rate is higher than either the As-Was or the As-Is environment. The results across all three risk frameworks is summarized in Fig. 12.13, and clearly shows that implementing the STARS filter will not reduce the number of accidents but will instead increase their frequency even further.

To understand why a proposed change that is intended to improve safety risk increases the number of expected accidents, especially in light of the realization that all of the objectives of the proposed change are realized, additional pathways through the risk framework must be evaluated. The conditional probability of *No Alert* given a true conflict in the To-Be environment increases substantially over the As-Is from 0.0067 to 0.011. This data reflects the likelihood of a real aircraft not being recognized by the RADAR due to the de-sensitivity being implemented through the new filtering functionality. One of the reasons why intuition fails to correctly estimate this result is that in the To-Be environment, the conditional probability of an intervention given no alert, and the conditional probability of an accident for both the intervention and no intervention cases, remains the same as it was in the As-Is case. However, the risk is sensitive to the number of missed alerts, and the increase in this value leads to an increase in the joint probability of an accident given no alert when there is a true conflict. This increase exceeds the reduction achieved by the reduction in false alerts and restoration of controller confidence. Figure 12.14 shows the As-Is and To-Be risk frameworks side-by-side with the relevant values highlighted for easier examination.

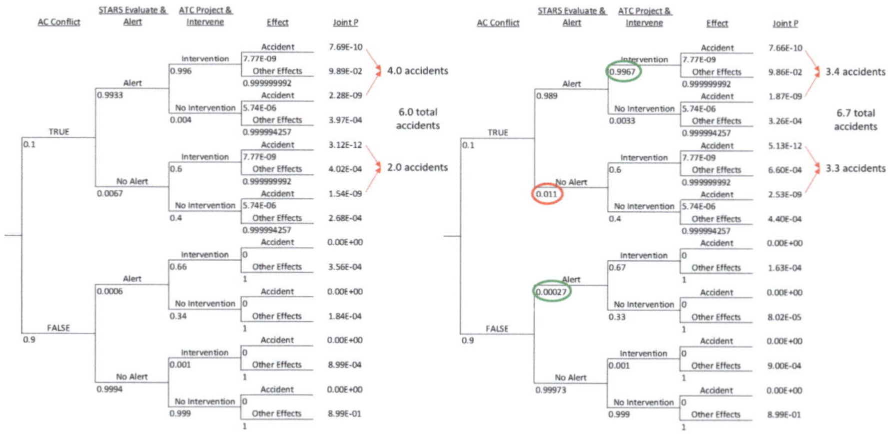

Fig. 12.14 Comparison of As-Is and To-Be frameworks

12.10 Summary

This case study demonstrates the power of the Human Viewpoint methodology in a typical engineering management application. Perhaps the most important insight gained from this case study is that managers cannot rely on intuition to serve as the basis for risk-based decisions. As a superior alternative to intuition, the Human Viewpoint methodology provides a platform from which a variety of data-driven, quantitative tools can be applied to gain a more complete understanding of a proposed change's influence on the performance of a complex system. Only with this level of understanding can engineering management professionals harbor a reasonable expectation that their decisions will consistently add value to their enterprise.

Chapter 13
Human Performance Modeling for Distracted Driving

Cansu Kandemir
cansuknd@gmail.com

Abstract This chapter develops a human performance simulation model in the context of distracted driving. The driver's cognitive demand is quantified using mental workload methods that can capture the additional demand when nomadic devices, such as cell phones, are used while driving and impact performance. Additionally, the cognitive demand of driver's aids, such as lane departure warnings and blind spot warnings, can be included in the model. These driver's aids may improve driver performance but also incur an additional mental workload. The human performance model can be used to conduct multiple simulations with different combinations of devices in order to predict the impact on driver performance.

Keywords Human performance modeling · Mental workload · Multiple resource theory

13.1 Introduction

Human performance modeling is a method for quantifying, predicting and examining ways to improve human system interactions. This chapter focuses on measuring human performance by first identifying the cognitive demands associated with tasks and then using computational models to explore the impact of changing task demands on operator outcomes. It describes the process of using a combination of mental workload assessment and simulation modeling to investigate changes in human performance. Distracted driving is a well-known phenomenon that impacts driver safety. An advanced human performance simulation model that investigates the impact of nomadic devices and driver's aids on driver mental workload is described. The model establishes the use of mental workload as a surrogate for distracted driving.

H. A. H. Handley, *The Human Viewpoint for System Architectures*, Topics in Safety, Risk, Reliability and Quality 35, https://doi.org/10.1007/978-3-030-11629-3_13

13.2 Quantifying Mental Workload

The cognitive demands of operators interacting with systems to complete tasks can be explained through the use of mental workload. Hart and Staveland (1988) define mental workload as the connection between the amount of mental processing resources available and the amount required by the task. Mental workload measures identify occurrences of increased task demands resulting in low performance. Increase in task demands can cause overloads which results in inefficient processes and poor performance along with ergonomic and mental health symptoms (Lindblom and Thorvald 2014). Modifying task demands to avoid overloading the operator can increase safety, comfort and performance.

The main objective of measuring mental workload is to quantify the mental cost of performing tasks in order to predict operator and system performance (Cain 2007). Mental Resource Theory (MRT) provides an analytical method to understand the relationship between mental resources and task demands while multi-tasking in a complex environment (Wickens 2002). According to MRT, the human mind can assign visual, auditory, cognitive, motor, and speech resources to task demands, either individually or jointly until demands exceeds available resources. In case of multi-tasking, when task demands intersect, less resources are available to assign to accomplish each task. As task demands increases, performance level usually drops, response times and errors increase, control variability increases, fewer tasks are completed per unit time, and task performance strategies change (Huey and Wickens 1993).

There are different measurements techniques for mental workload, including psychophysiological, subjective, and performance measurement techniques (Miller 2001). Psychophysiological measurement of workload is a concept based on evidence that increased mental demands lead to increased physical responses from the body. Psychophysiological workload measures rely on continuous measurement of the physical responses of the body using sensors. Subjective measurement is based on the use of rankings or scales to measure the amount of workload a person is feeling. Subjective workload measures rely on question-answer type responses to varying levels of workload. Performance measurement of workload relies on examining the capacity of an individual by means of a primary or secondary task. An estimate of mental workload can be determined by measuring how well a person performs on the task, or how their performance worsens as workload increases. A summary table of the mental workload measurement techniques can be found in Table 13.1.

System designers can use various methods, such as rating scales, questionnaires, or interviews, to gather subjective workload data. Hart and Wickens (1990) subdivided rating scale methods into one-dimensional and multi-dimensional ratings. One-dimensional ratings are easy to understand and use but are considered too simple to measure the complexity of workload. They lack the ability to combine ratings for predicting workload in different situations that involve similar tasks. While one-dimensional measures are more sensitive, multi-dimensional measures are more diagnostic. Moreover, most of the multi-dimensional scales, such as

Table 13.1 Summary of workload measures (Miller 2001)

Workload measures		
Physiological measures	Subjective measures	Performance measures
1. Cardiac	1. One-dimensional scales	1. Primary task performance
2. Respiratory	1.1 Modified Cooper-Harper scale	2. Secondary task performance
3. Eye	1.2 Overall workload scale	
4. Speech measures	2. Multi-dimensional scales	
5. Brain activity	2.1 NASA task load index scale	
	2.2 Subjective workload assessment technique	
	2.3 Visual, auditory, cognitive, psychomotor method	

the visual, auditory, cognitive and psychomotor (VACP) model have a predictive capability through constructive modeling (McCracken and Aldrich 1984). These techniques can be used to predict mental workload when the system is just a concept and no prototype exists. Any task performed by an operator can be broken down into these components and rating scales can provide a relative rating of the degree to which each resource component is used.

Subjective workload measures that are used in analytical modeling, such as VACP, focus on task demand in multiple resources. When those measures are used in simulations that includes the task duration, it produces aggregate measures that are sensitive to both task difficulty and time. When combined with a detailed task analysis, simulation models provide a predictive analysis of human performance (Wickens 2002). Some simulation software, such as the Improved Performance Research Integration Tool (IMPRINT), has a mental workload scale imbedded in the tool. The VACP scale used in IMPRINT is shown in Table 13.2. IMPRINT adjusts the task completion parameters based on the identified user interfaces that determine the workload channels (Mitchell and Samms 2009).

13.3 Simulation Tools for Modeling Workload

Simulation models help system designers to predict task execution and the corresponding mental workload levels. Simulation models capture the task decomposition to accomplish a particular work process, the amount of time and mental resources it takes to execute each task, the sequence of the tasks in the process, and the assigned operator for each task. The time and effort needed to identify and quantify the model artifacts (e.g. tasks, operators, time, resources and interfaces) to design the resulting model can be extensive. Subjective methods are the most often used methods to evaluate the workload associated with a task. These methods, especially those with rating

Table 13.2 VACP workload estimation scales (Mitchell and Samms 2009)

Workload demand value
Visual
3.0—Visually register/detect
3.0—Visually inspect/check
4.0—Visually locate/align
4.4—Visually track/follow
5.0—Visually discriminate
6.0—Visually scan/search/monitor
5.1—Visually read
Auditory
1.0—Detect/Register sound
2.0—Orient to sound (general)
4.2—Orient to sound (selective)
4.3—Verify auditory feedback
3.0—Interpret semantic content (speech) simple (1–2 words)
6.0—Interpret semantic content (speech) complex (sentence)
6.6—Discriminate sound characteristics
7.0—Interpret sound patterns
Cognitive
1.0—Automatic (simple association) all values below 7.0 map to
1.2—Alternative selection solving
3.7—Sign/Signal recognition
4.6—Evaluation/Judgment (single aspect)
5.0—Rehearsal
5.3—Encoding/Decoding, recall
6.8—Evaluation/Judgment (several aspects)
7.0—Estimation, calculation, conversion
Psychomotor
2.2—Discrete actuation (button, toggle trigger)
2.6—Continuous adjustable (flight control, sensor control)
4.6—Manual (tracking) fine motor discrete
5.5—Discrete adjustable (rotary, vertical thumb wheel, lever position)
6.5—Symbolic production (writing)
7.0—Serial discrete manipulation (keyboard entries)

scales, have various advantages for measuring workload relative to other approaches, i.e., they have good face validity and general applicability. The VACP method, based on MRT, is one of the preferred methods. According to MRT, a human has several different types of resources; these resources are distinguished by information processing stages (encoding and central processing or responding), perceptual modality (auditory or visual), and processing codes (spatial or verbal) (Wickens 2002). Tasks can be configured with workload channels defined to correspond to these diverse dimensions of MRT.

Simulation models created to evaluate the impact of mental workload on performance often use a task network approach. Task network models allow the performance of individual operators to be analyzed by decomposing the operator's task assignments into a series of main tasks and then into series of sub-tasks. Furthermore, human performance simulation modeling has the capability to include the effects of the operator's education, experience, and workplace conditions to include in the mental workload analysis (Mitchell and Samms 2009). While IMPRINT is one of the more popular workload simulation software, other workload simulation modelling tools are available, such as the Integrated Performance Modeling Environment (IPME) (Dahn and Laughery 1997). Generally, a mental workload modelling capability can be integrated with any Discrete Event Simulation (DES) tool to predict the operator's workload, however, it will require the modeler to create task level system interfaces in order to conduct MRT type analyses.

13.4 A Driver Workload Simulation Model

In general, the application of human performance modeling with computational mental workload is seen in military and health-care environments for critical processes that require immediate attention and decision-making. However, the popularity and application areas of human performance models is growing rapidly. Driver interface design (Kandemir et al. 2018), manufacturing systems analysis (Bommer and Fendley 2016), and various process control applications (Kandemir and Handley 2018) are some of the areas that are developing computational human performance models.

VACP workload method is used to develop the driver workload simulation model. The task network for the driver workload model was created using the NHTSA guidelines (NHTSA 2013) and the Federal Motor Carrier Safety Administration (FMCSA) scenario (Robin et al. 2005). The driving scenario used in the model is 13 miles long and includes two lane roads, four lane divided highways, and four lanes non-divided highways with up to 55 mph speed limit of light to moderate traffic.

First, the driving scenario is broken down into driving segment functions and tasks that coincide to driving changes, either by roadway or intersections, then, driver's interfaces related with each driving segment is determined, and lastly, the workload data that corresponds to each driving segment and interface is entered (Kandemir et al. 2018). For large vehicles, the Army Research Laboratory (ARL) has identified categories of driving functions that are shown in Table 13.3 (Mitchell

Table 13.3 ARL driver related functions (Mitchell 2009)

Function	Function
Driver awareness	Assess vehicle orientation
	Assess vehicle motion
	Assess vehicle function
	Assess vehicle traction
Driver visualizations	Recognize path
	Determines distance to objective
	Scan sector
Driver manipulations	Accelerate
	Decelerate
	Coast

2009; Wojciechowski 2004). Moreover, NHTSA has identified some categories of truck driving related activities (NHTSA 2013). These two resources are used to create the main driving function-task decompositions of the model. A set of distracters and driving aids were added to this base scenario model for attaining the necessary workload analysis.

The model was created in the IMPRINT tool by capturing the scenario as a task network. Figure 13.1 depicts the top level activity diagram of the model. For each activity shown, a sub model exists that further decomposes that function into scenario segments. For each segment, subtasks were created that contain the specific steering tasks for that segment. Throughout the scenario the situation awareness function includes subtasks that capture the visual scanning tasks that the driver performs to monitor the environment. The maintain speed subtasks are used by each scenario segment to maintain speed, accelerate, slow down and stop.

Once the function-task network is completed, the workload data is entered using IMPRINT's VACP feature that links human resources (visual, auditory, cognitive and psychomotor) to interfaces. Interfaces available for the driver in the base scenario are shown in Table 13.4. The 7-point VACP scale that is used to assign workload values is shown in Table 13.2. A sample task-interface matching and workload value assignment can be found in Table 13.5. For instance, the task "Check warning sign" in "Pass rail road crossing" function requires use of the "Windshield" interface. It has the associated resources of visual (align the car in the lane) and cognitive (evaluate the traffic light). The cognitive workload is set to 4.6 which represents the mental workload of "evaluation/judgement of a single aspect" and the visual workload is set to 4.0 which represents the mental workload of "visually locate/align". Workload values used in the simulation were obtained from ARL studies to ensure the validity. Based on these inputs, IMPRINT's workload algorithm calculates the mental workload throughout the driving scenario, i.e., the IMPRINT software sums the workload ratings for each resource across concurrent tasks. While each resource can be assigned to workload value of 7 at most, the operator workload can be higher than 7 since the IMPRINT's algorithm sums the workload in multi-tasking occurrences.

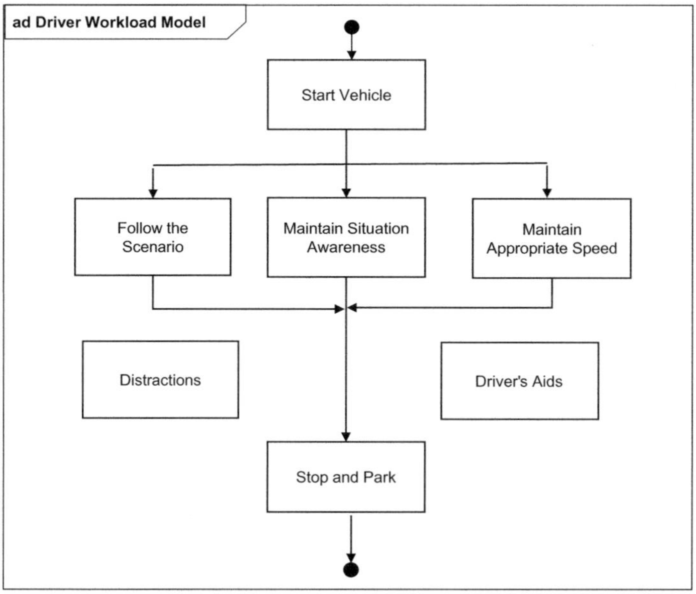

Fig. 13.1 Top level of activity diagram of the driver workload model

Table 13.4 Driving interfaces (Kandemir et al. 2018)

#	Name
1	Steering mechanism
2	Instrument panel/dashboard
3	Accelerator
4	Brake
5	Driving display
6	Windshield

Table 13.5 Workload data example for "Pass rail road crossing"

Function	Task	Interface	V	A	C	P
Pass rail road crossing	Go straight	Steering mechanism	0	0	1.0	4.6
	Check warning sign	Windshield	4.0	0	4.6	0
	Slow down	Accelerator	0	0	4.6	2.6
	Stop	Brake	0	0	1.2	2.2
	Stop and wait	Windshield	3.0	0	1.0	0
	Accelerate	Accelerator	0	0	4.6	2.6
	Continue	Steering mechanism	0	0	1.0	2.6

13.5 Including Distracters and Drivers Aids

There are two categories of distracters as determined by NHTSA (2013): integrated distractors which are the existing interfaces in the vehicle and nomadic distractors which are the devices brought into the vehicle by driver. Cruise control and climate control are an example of integrated distracters. Cell phones and non-integrated GPS are an examples of nomadic distracters. Using these interfaces while driving increases the likelihood of multi-tasking that results in high level of mental workload for the driver.

Cell phone and Global Positioning System (GPS) devices are added to the base model as the nomadic distracters. Cell phone adds two main functions to the model: text message and voice call. GPS device adds one main function to the model: follow the directions. These main functions are decomposed into sub-tasks.

In the simulation model, all the distractors happen randomly. Once a distractor occurs, the driver must continue driving and complete the distractors tasks (i.e. respond to the call, check GPS). Interfaces related with the distractors are defined and associated VACP workload values are entered. The scenario of the distractors included to the simulation model are as follows:

Cell Phone: Randomly after 900–1200 s of the simulation scenario the call starts. The driver listens and responds between randomly 5–60 s. Figure 13.2 shows the model implementation of "voice call". Randomly after 300–900 s of the simulation scenario the text-based communication starts. The driver reads, types and waits for the response randomly between 5 and 60 s each for a random number of cycles. The model implementation of a text-based communication is shown in Fig. 13.3.

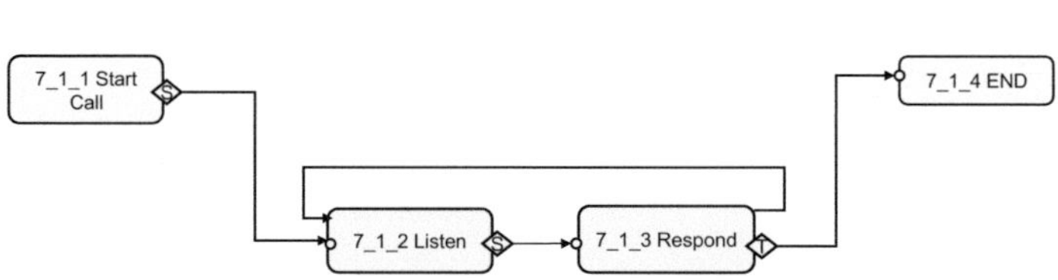

Fig. 13.2 Model implementation of a voice call

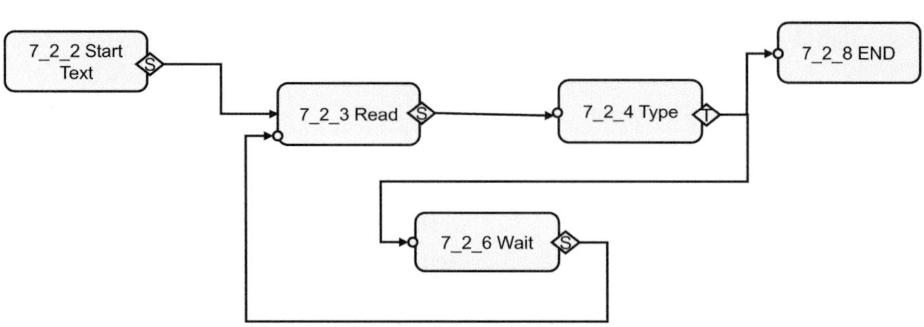

Fig. 13.3 Model implementation of text based communications

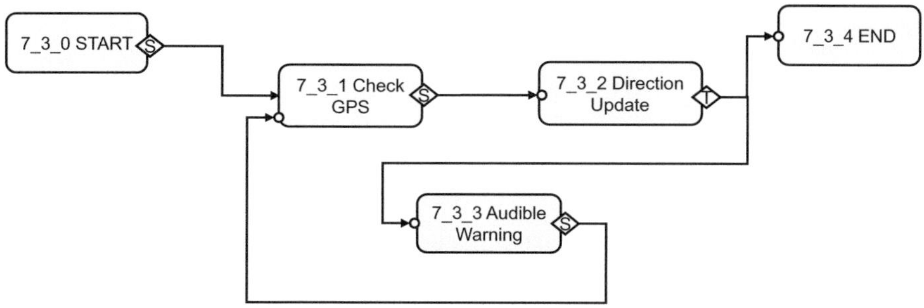

Fig. 13.4 Model implementation of the GPS

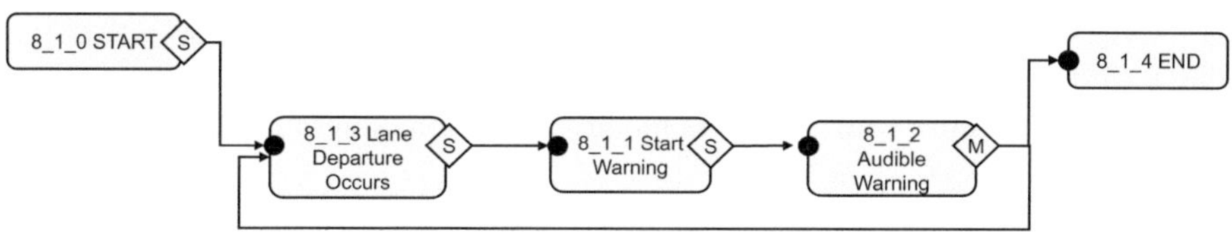

Fig. 13.5 Model implementation of the lane departure warning

GPS: The GPS scenario starts with the start of the driving tasks. The driver checks the GPS randomly approximately for 3–5 s while continuing driving. Moreover, GPS gives audible direction warning before 5 s of turning and merging for about duration of 5–6 s. Figure 13.4 shows the model implementation of "follow the directions".

Lane Departure Warning (LDW) and Blind Spot Warning (BSW) are added to the base model as the driver's aid systems. For the driver's aids analysis, similar to the distractors, the interfaces related with the driver's aids are defined and associated VACP workload values are entered. Note that some driver aids change the way the driver operates the vehicle (e.g. BSW), while others change the focus of the driver's attention (e.g. LDW). The scenario of the driver's aid included with the simulation model are as follows:

LDW: Lane departure warnings are on when the turn signal is "not on". LDW warns the driver when the vehicle begins to move out of its lane. The aim of these systems is to minimize accidents caused by leave-the-lane collisions. The lane departure in the simulation scenario occurs randomly while the driver is driving straight. It gives an audible warning that stops other non-driving tasks for 20 s while performing additional situation assessment. It is assumed that the warning is effective in provoking a necessary response from the driver. Figure 13.5 shows the model implementation of the LDW.

BSW: Blind spot warnings warn the driver if another vehicle is in the next lane and slightly behind the driver, usually in the driver's "blind spot". These systems are designed to prevent accidents caused by absence of situation awareness by the driver of other vehicles in the sideline. BSW system will have the effect of relieving the driver of having to scan for vehicles in the adjacent lane. As a result, BSW lowers

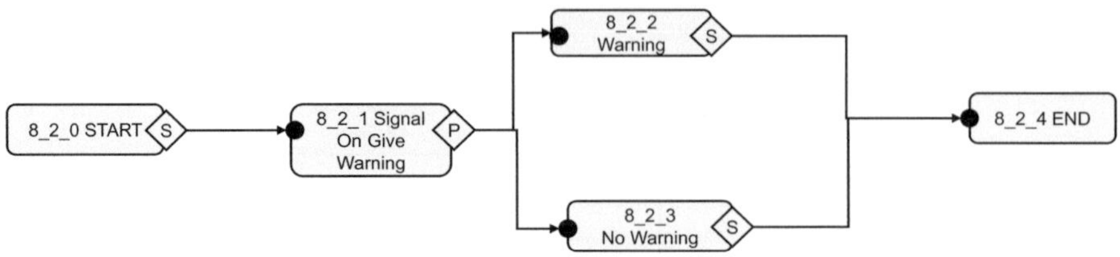

Fig. 13.6 Model implementation of the blind spot warning indicator

the driver's visual workload while turning, merging, and changing lanes. If there's a vehicle in the blind spot while turning, merging, and changing lane an audible warning is given. In the simulation scenario, there is a 50% chance that there is a vehicle at the blind spot. The driver stops other non-driving tasks for 20 s to attend to the warning. Figure 13.6 shows the model implementation of the BSW.

13.6 Simulation Results

The workload output reports generated by IMPRINT were analyzed to assess the predictions and practicality of the model. The main findings on the initial baseline driving simulation, the driving simulation with distracters, and the simulation with driver aids are explained in this section.

In a workload simulation study, in order to make meaningful predictions, the nominal values of workload must be determined, and the workload threshold must be defined. The workload threshold is used to indicate the overloaded conditions. It is important to note that threshold levels are defined for particular scenarios; the specific value is not as important as differences from the baseline value in the subsequent simulations. A baseline simulation, with no distracters or driver aids included, was conducted to determine nominal values of workload and the workload threshold across the driving scenario. These values were used to evaluate deviations from the baseline values when distractors and driver's aids were added to the driving scenario. Table 13.6 shows a sample of the results for the baseline case of "Merging/Changing Lanes." Based on the results of the entire baseline simulation, the workload threshold for this study can be defined as 40. This is the value where the operator is fully engaged, and it is anticipated that additional workload beyond this level will lower performance and cause errors.

The next configuration of the driver workload model included both texts and cell calls (nomadic distracters). Table 13.7 shows a similar portion of the scenario while listening to a cell phone call. The simulation outcomes identify the tasks that are contributing to the high workload during the texting and the voice portion of the scenario. Note the increase in workload values.

The next nomadic distractor included in the model is GPS. Table 13.8 shows a similar portion of the scenario while the driver is checking the GPS directions. The

Table 13.6 Baseline model workload outcomes example

Segment	Clock	Function name	Task name	Overall workload
Merging/Changing lane	00:28:49.58	Situation awareness—monitoring	Maintain direction control (while driving straight)	25.60
		Situation awareness—monitoring	Monitor while merging/turning	
		Maintain speed	Slow down	
		Right turn	Slow down	
	00:28:55.98	Maintain speed	Maintain speed (while driving straight)	23.60
		Situation awareness—monitoring	Monitor while merging/turning	
		Right turn	Turn the wheel right	
	00:29:01.16	Right turn	Accelerate	25.60
		Maintain speed	Accelerate	
		Situation awareness—monitoring	Maintain direction control (while driving straight)	
		Situation awareness—monitoring	Monitor while merging/turning	

simulation outcomes identify the tasks that are contributing to the high workload during the checking the GPS directions and listening to the GPS direction warnings portion of the scenario. There is a substantial increase in workload values.

Table 13.9 shows the portion of the scenario when two tasks of GPS overlap while driving straight. Even though it is not very likely to happen, this alone can increase the driver's workload substantially. Table 13.10 shows the portion of the scenario when two different distractor tasks overlap; namely talking on the phone and checking the GPS directions. When two distractors occur at the same time, with the increase in the workload, the tasks may not be manageable by the driver.

The last configuration of the driver workload model includes the driver aids. Table 13.11 includes the workload of the driver while using LDW and BSW systems for a comparable function. As shown in the table, these systems increase the awareness of the driver and do not increase or decrease the workload of basic driving tasks in a significant amount.

This chapter summarizes the main steps of developing a discrete event workload simulation for driver system designers and/or driving related policy makers. For the in-depth analysis of the distractors and driver's aid, readers are referred to Kandemir

Table 13.7 Baseline model with text and call workload outcomes example

Segment	Clock	Function name	Task name	Overall workload
Listen while turning/merging	00:25:12.98	Situation awareness—monitoring	Monitor while merging/turning	38.00
		Call	Respond	
		Merge/Slight right left	Slow down	
		Maintain speed	Slow down	
		Merge/Slight right left	Turn the wheel slowly	
	00:25:28.00	Call	Listen	40.00
		Situation awareness—monitoring	Monitor while merging/turning	
		Merge/Slight right left	Slow down	
		Maintain speed	Slow down	
		Merge/Slight right left	Turn the wheel slowly	
	00:25:45.92	Call	Listen	38.40
		Situation awareness—monitoring	Maintain direction control (while driving straight)	
		Situation awareness—monitoring	Monitoring driving straight (forward, mirrors, dashboard)	
		Merge/Slight right left	Slow down	
		Maintain speed	Slow down	

et al. (2018). The results from the driver workload simulation model can be used to identify problematic areas related to driver multi-tasking caused by using distracters or driver aid systems. By changing the number and type of interfaces that the driver must manage, the causes of driving workload and overload can be explored. After the interface is identified as a cause of driver distraction, system designers can determine ways to reduce the workload associated with these secondary tasks through better device interface design, integration of the device with existing interfaces and/or driver enforcement policies.

Table 13.8 Baseline model with GPS workload outcomes example

Segment	Clock	Function name	Task name	Overall workload
Merging/Changing lane	00:28:49.58	Situation awareness—monitoring	Maintain direction control (while driving straight)	35.40
		Situation awareness—monitoring	Monitor while merging/turning	
		Maintain speed	Slow down	
		GPS	Monitor GPS	
		Right turn	Slown down	
	00:28:55.98	Maintain speed	Maintain speed (while driving straight)	33.40
		GPS	Monitor GPS	
		Situation awareness—monitoring	Monitor while merging/turning	
		Right turn	Turn the wheel right	
	00:29:01.16	Right turn	Accelerate	25.60
		Maintain speed	Accelerate	
		Situation awareness—monitoring	Maintain direction control (while driving straight)	
		Situation awareness—monitoring	Monitor while merging/turning	

Table 13.9 Baseline model with GPS workload outcomes example when two GPS tasks overlaps

Segment	Clock	Function name	Task name	Overall workload
Go straight	00:01:10.00	Straight with traffic light	Go straight	43.10
		Situation awareness—monitoring	Maintain direction control (while driving straight)	
		GPS	Monitor GPS for directions	
		GPS	Audible direction warning	
		Maintain speed	Maintain speed (while driving straight)	
		Situation awareness—monitoring	Monitoring driving straight (forward, mirrors, dashboard)	

Table 13.10 Baseline model with call and GPS workload outcomes example

Segment	Clock	Function name	Task name	Overall workload
Talk while merging/turning	00:18:12.00	Maintain speed	Maintain speed (while driving straight)	44.20
		Situation awareness—monitoring	Monitor while merging/turning	
		GPS	Monitor GPS	
		Call	Respond	
		Merge/Slight right left1	Turn the wheel slowly	
	00:18:45.47	Situation awareness—monitoring	Maintain direction control (while driving straight)	46.20
		Situation awareness—monitoring	Monitor while merging/turning	
		Call	Respond	
		GPS	Monitor GPS	
		Maintain speed	Slow down	
		Right turn	Slown down	

Table 13.11 Baseline model LDW and BSW outcomes example

Segment	Clock	Function name	Task name	Overall workload
Merging/Turning with BSW	00:09:53.59	Situation awareness—monitoring	Maintain direction control (while driving straight)	20.40
		Situation awareness—monitoring	Monitor while merging/turning	
		Maintain speed	Slow down	
		Right turn	Slow down	
	00:09:57.27	Maintain speed	Maintain speed (while driving straight)	19.40
		Situation awareness—monitoring	Monitor while merging/turning	
		Right turn	Turn the wheel right	
		Blind spot warning	Warning	
	00:10:02.13	Right turn	Accelerate	21.40
		Maintain speed	Accelerate	

(continued)

Table 13.11 (continued)

Segment	Clock	Function name	Task name	Overall workload
		Situation awareness—monitoring	Maintain direction control (while driving straight)	
		Situation awareness—monitoring	Monitor while merging/turning	
		Blind spot warning	Warning	
	00:10:03.25	Right turn	Accelerate	20.40
		Maintain speed	Accelerate	
		Situation awareness—monitoring	Maintain direction control (while driving straight)	
		Situation awareness—monitoring	Monitor while merging/turning	

13.7 Summary

Human performance models that include mental workload can help human system engineers understand the complex phenomena of human performance, while saving significant amount of time and cost in analyzing prototype systems and running experiments (Kandemir and Handley 2018). In this chapter, the well-known methods of quantifying, measuring and analyzing human performance using mental workload and computational modeling are summarized. Using this approach, human performance can be estimated using a simulation model that includes the impact of workload on operator outcomes. The use of MRT and the VACP scale of human resource components allows a simulation model to provide quantitative predictions of workload during system development. The driver workload model explained in this chapter illustrates the ability to identify causes of high workload due to additional interfaces that task the driver's mental resources. The resulting model can be used to evaluate system changes designed to ameliorate the high workload and improve the human performance.

References

Bommer SC, Fendley M (2016) A theoretical framework for evaluating mental workload resources in human systems design for manufacturing operations. Int J Ind Ergon. https://doi.org/10.1016/j.ergon.2016.10.007

Cain B (2007) A review of the mental workload literature. Technical report, Defence Research and Development Canada, Toronto

Dahn D, Laughery K (1997) The integrated performance modeling environment; simulating human-system performance. In: Proceedings of the 29th winter conference on simulation, Atlanta, GA

Hart S, Staveland L (1988) Development of NASA-TLX (task load index): results of empirical and theoretical research. Adv Psychol 52:139–183

Hart S, Wickens C (1990) Workload assessment and prediction. In: Booher H (ed) MANPRINT. Springer, Netherlands, pp 257–296

Huey B, Wickens C (1993) Workload transition: implications for individual and team performance. The National Academies Press

Kandemir C, Handley H (2018) Work process improvement through simulation optimization of task assignment and mental workload. Comput Math Org Theory

Kandemir C, Handley H, Thompson D (2018) A workload model to evaluate distracters and driver's aid. Int J Ind Ergon 63:18–36

Lindblom J, Thorvald P (2014) Towards a framework for reducing cognitive load in manufacturing personnel. In: Proceedings of the applied human factors and ergonomics conference (AHFE)

McCracken J, Aldrich T (1984) Analyses of selected LHX mission functions: implications for operator workload and system automation goals. Technical report, U.S. Army Research Institute Aviation Research and Development Activity, Fort Rucker, AL

Miller S (2001) Literature review: workload measures. The University of Iowa, Iowa City

Mitchell D (2009) Successfully changing conceptual system designs using human performance modeling. In: Presented at the human system integration symposium, Annapolis, MD

Mitchell D, Samms C (2009) Workload warriors: lessons learned from a decade of mental workload prediction using human performance modeling. In: Proceedings of the human factors and ergonomics society annual meeting, vol 53, no 12, pp 819–823

NHTSA (2013) Visual-manual NHTSA driver distraction guidelines for in-vehicle electronic devices. National Highway Traffic Safety Administration

Robin L, Knipling R, Tidwell S et al (2005) FMCSA commercial truck simulation validation study phase I pilot test: driving scenario definition and development. In: Paper presented at the driving simulator conference—North America, Orlando, FL

Wickens C (2002) Multiple resources and performance prediction. Theor Iss Ergon Sci 3(2):159–177

Wojciechowski J (2004) Validation of improved research integration tool (IMPRINT) driving model for workload analysis. Technical report, Aberdeen Proving Ground, MD

Chapter 14
A Human View Approach to Risk Management

C. Ariel Pinto and Mark Guilford
cpinto@odu.edu

Abstract This chapter provides a set of operational risk management questions that can be used to identify, analyze and adjudicate risk events. The questions are applied within a system context by identifying an operational scenario and focusing on the impact of an event to the set of system goals. The risk management questions can be aligned with the stages of the Human Viewpoint methodology and evaluated using the human performance metrics of the socio-technical system. An example of the risk management strategy is provided in the context of identifying and mitigating the human role in a phishing risk scenario.

Keywords Operational risk · Risk management · Risk events

14.1 Introduction

In the context of systems engineering, an operation can be described as a set of processes designed to attain the objectives of the overall system. Risk, on the other hand, is often thought of as the impact to the operation when events occur that cause the process to fail to achieve its objectives. As described in earlier chapters, the Human Viewpoint methodology provides an approach to describe a socio-technical system using a set of models; this framework can be used to identify and mitigate operational risks due to the human component of a socio-technical system. As an example, the Human Viewpoint methodology is applied to identify the operational risk from "phishing," i.e. the act of deceiving individuals into giving up personal information. Using risk management techniques, this human-based risk can be identified in various failure scenarios to evaluate the ability of the system to achieve its stated goals.

H. A. H. Handley, *The Human Viewpoint for System Architectures*, Topics in Safety, Risk, Reliability and Quality 35, https://doi.org/10.1007/978-3-030-11629-3_14

14.2 Operational Risk Management Questions

Operational risk can be qualitatively described as the potential undesirable consequences directly or indirectly resulting from failure of one or more elements of the system of interest (Basel II 2004). An accident is an event that is both unintended and undesirable; hazards, objects, actions, processes, or conditions may contribute towards the occurrence of an accident. Risk is an event with undesirable consequences without specific regards to intent. Kaplan (1997) quantitatively described risk as a function of a set of risk scenarios, the likelihood of the scenarios, and the damage as a result of the scenario consequences.

Risk management is a formal process used to continuously identify, analyze, and adjudicate risk events (Garvey 2008). There are many risk management processes used in various industries, disciplines, and professions. However, there are seven generalizable guiding questions in risk management that encompass most of these processes:

1. What should go right?
2. What can go wrong?
3. What are the causes and consequences?
4. What is the likelihood of occurrence?
5. What can be done to detect, control, and manage them?
6. What are the alternatives?
7. What are the effects beyond this particular time?

A system is defined not merely by the enumeration of its subsystems or elements, but more importantly by articulation of its goals (or dually, its constraints). These goals or constraints, i.e., *What should go right?*, trigger engineering and management endeavors and are basic aspects of any design process. Although it may seem trivial, the underlying principle in this first step is that in order to know what can go wrong, one must first know what should go right.

After the ideal or right scenarios have been articulated in the form of objectives and constraints, one can now proceed with identifying *What can go wrong?* For the most part, identifying risk events is primarily done by looking back on what has gone wrong in the past and the knowledge of processes resulting in events other than those desired, i.e. those that are supposed to go right. Negative scenario identification is one common strategy that basically considers different ways things can go wrong in a system based on what are the known desired events. It is often helpful to imagine departures or deviations from the ideal. Using natural language, i.e., affixing "not" to statements of objectives and constraints will form first order—albeit very simple—risk statements. This has been previously referred to as the anti-goal by Pinto et al. (2010).

Once risk events are identified, the next phase is to describe these events for the purpose of extending the understanding and knowledge about the event. This involves establishing causality, identifying root causes and their likelihood, as well as characterizing consequences and impact, i.e., *What are the causes and consequences?* This helps in developing more appropriate and effective decisions or actions related to the management of risk. Establishment of causes and consequences is founded on the evidentiary relationship between events such that the occurrence of one event implies the occurrence of the other. However, the strength of this causal relationship may depend on the details of their relationships of necessity and sufficiency. Necessary cause relationship suggests that a set of events (e.g., set B) is described to be necessary to cause another set of events (e.g., set A) if B is a required condition for the occurrence of A, not that A actually occurs. On the other hand, sufficient cause relationship suggests that a set of events (e.g., set B) is described to be sufficient to cause another set of events (e.g., set A) if the occurrence of B guarantees the occurrence of A.

Sequences of events that lead to a particular risk event, that is, a causal chain of events, need to be described in terms of their respective chances of occurrence. *What is the likelihood of occurrence?* captures the frequency or chance of occurrence of a risk event as a quantitative or qualitative description of how often or how soon a particular risk event may occur. This is often derived from historical information or records of the risk event. This can also be derived from team based elicitation. The notions of necessary-and-sufficient causes taken together form the foundation of judging causality in many fields, including systems engineering and risk analysis. From a risk management perspective, the ultimate (albeit possibly impossible) goal is to identify the necessary-and-sufficient set of causes, where an event B is necessary and sufficient condition for another event A if A occurs if-and-only-if B occurs. That is, $P(A|B) = 1$ and $P(A'|B') = 1$.

Ranking and scoring is conducted to evaluate criticality and determine relative importance of risks, i.e. *What can be done to detect, control, and manage them?* What may be critical is the context of the risk. Common critical risks are those whose consequences are related to health and safety, compliance to regulatory requirements, or those that affect core mission and operational objectives. Criticality may be assessed using a risk matrix similar to that shown in Fig. 14.1. This risk matrix highlights risk events with high severity ratings, such as those risks that fall under the catastrophic category of consequences or risks that fall under the very likely category of likelihood of occurrence. However, particular attention should be given to those risks in which consequences are catastrophic and the likelihood of occurring is very likely or eminent. In Fig. 14.1, these are the risk events that fall in the darker boxes. Risk events that fall in the darkest boxes should be addressed immediately. Risk matrix tables are useful for categorizing and prioritizing identified risks.

The last question determines which risk treatment strategies will work well together, given the causal chain of events, i.e., *What are the alternatives?* Risk treatment strategies are not mutually exclusive, and effective action plans are usually made up of combination of strategies, albeit in various degrees. In general, risk treatment strategies are identified for reducing chances of occurrence, for reducing consequences if they do occur, or both. Detection and control are the typical strategies to reduce the chances of occurrence and are often applied in anticipation of a risk event, while recovery plans address the reduction of consequences after risk events have occurred.

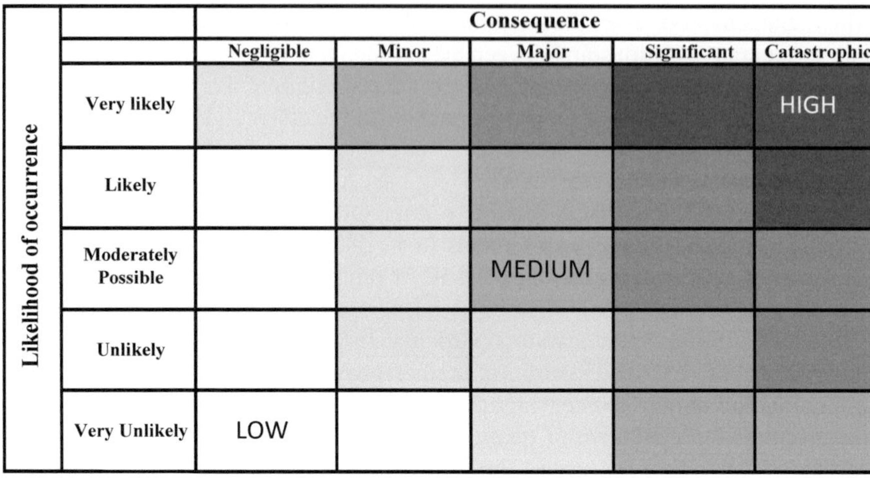

Fig. 14.1 Common risk matrix with consequence and likelihood ratings

From a system perspective, it is important to evaluate the effects of the risk treatment alternatives on other elements of the system. Risk treatment alternatives may be analyzed according to their effects on the functionality of other elements, the manner by which the risks alter the interaction among elements, and the potential of risks to affect future decisions, i.e., *What are the effects beyond this particular time?* This is also the point where the acceptable risk levels are determined by comparing the costs and benefits of each mitigation alternative. The concept of "As Low as Reasonably Practicable", a fundamental approach that sets the risk to the tolerable, reasonable, and practical level, is an example of a risk approach for this phase. There is also the notion of residual and emerging risks, which are manifestations of the fact that no risk events can be totally eliminated and that new ones may emerge in the process of mitigating others.

14.3 The Human Viewpoint Aligned to Risk Management

As described in earlier chapters, the Human Viewpoint methodology provides a process to describe the human system and capture it in a set of models to augment the system architecture description. It consists of a sequence of five iterative steps: Context, Data, Models, Analysis, and Fit for Purpose (Handley and Knapp 2014). These steps can provide a framework for addressing the seven generalizable risk management guiding questions for a socio-technical system, as shown in Table 14.1.

An example of applying a risk management strategy in conjunction with the Human Viewpoint methodology can be illustrated by extending the communication system used as an example in the earlier chapters. Recall the communication system

Table 14.1 Risk management questions mapped to human viewpoint methodology

Risk management questions	Human viewpoint methodology
1. What should go right? 2. What can go wrong?	*Context step* identifies the scope of human focused data pertinent to the area of stakeholder concern
3. What are the causes and consequences? 4. What is the likelihood of occurrence? 5. What can be done to detect, control, and manage them?	*Data step* captures relevant attributes of each of the elements *Models step* illustrates the important relationships between the data elements that impact the system design
6. What are the alternatives? 7. What are the effects beyond this particular time?	*Analysis step* analyzes different use cases to provide analytic data to support the decisions consistent with the context *Fit for Purpose Views step* communicate results of analyses to support stakeholder decisions

Table 14.2 Communication system context stage risk management questions

Goals	What can go right?	What can go wrong?
Timeliness	Softphone calls should be initiated within 10 s	Softphone calls are not initiated within 10 s
Accuracy	Softphone calls should connect on the first try	Softphone calls do not connect on the first try
Availability	The softphone should be operational 95% of the time	The softphone is operational less than 95% of the time

is used to extend tactical radio networks and is installed on select vehicles (Handley et al. 2015). The area of stakeholder concern driving the Human View analysis is focused on evaluating whether the crew member in the vehicle seat that has access to the communication system can adequately operate the new equipment. The Context stage focuses on identifying the goals for this system, i.e., timeliness, accuracy, and availability. At the context stage, the *What can go right?* and *What can go wrong*? questions can be evaluated, as shown in Table 14.2.

Continuing the operational analyses, possible causes and consequences of risk events for the communication scenario are addressed at the Data and Models stages. Table 14.3 evaluates the *What are the Causes?*, *What are the Consequences?* and *What is the Likelihood?* questions. The contributing causes would be, individually and collectively, sufficient causes for the consequences, e.g. a preoccupied operator is sufficient for softphone calls to be not initiated within 10 s. The likelihood of these occurring is then estimated.

Table 14.3 Communication system data and model stages risk management questions

Goal	What are the causes?	What are the consequences?	What is the likelihood?
Softphone calls are not initiated within 10 s	Operator is pre-occupied	Actions delayed	Very likely
	Softphone device not accessible	Information delayed	Very likely
Softphone calls do not connect on the first try	Soft client application error	Information delayed	Very unlikely
	Internet network issue	Call attempt abandoned	Very unlikely
The softphone is operational less than 95% of the time	Software not updated	Operation fails	Unlikely
	Device is faulty	Operation proceeds intermittently	Unlikely

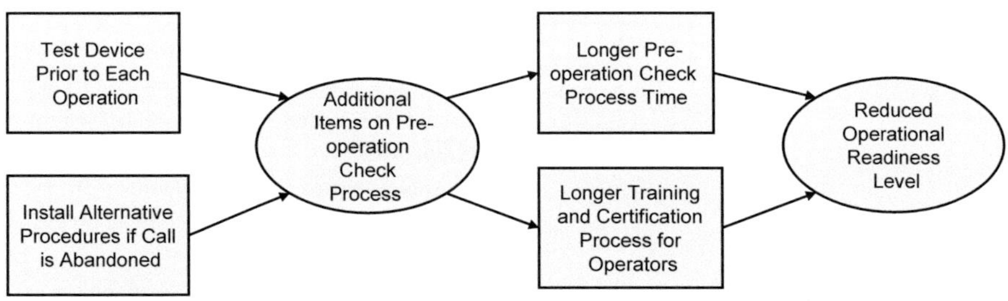

Fig. 14.2 Strategies to address operational risks with consequence for new risk event

At the Analysis and Fit for Purpose stages the last two risk management questions are addressed. Table 14.4 describes strategies, both preventative and consequence reduction, to address the *What are the alternatives?* question.

For the final question, *What are the effects beyond this particular time?*, consider the strategy "Test device prior to each operation" to prevent the contributing cause "Device is faulty", as well as the strategy "Install alternative procedures if call is abandoned" to reduce the consequence of "Call attempt abandonment". Both of these strategies may indeed reduce the overall operational risk but will create additional items on a pre-operation check process. This may in turn result in longer pre-operation check process time and longer training and certification process for operators due to the added procedure, as illustrated in Fig. 14.2. These two potential effects may then cause a reduction in overall operational readiness. The potential negative effects of the suggested amelioration strategies need to be considered as well as the potential reductions in risk.

Table 14.4 Communication system risk management strategies

Goal	Prevention strategies to address contributing causes		Consequence reduction strategies to minimize consequences	
	Cause	Strategy	Consequence	Strategy
Softphone calls are not initiated within 10 s	Operator is pre-occupied	Reexamine human task assignment	Actions delayed	Install operation delay procedures
	Softphone device not accessible	Move softphone closer to operator	Information delayed	Install operation delay procedures
Softphone calls do not connect on the first try	Soft client application error	Test application prior to use	Information delayed	Install alternative procedure if call abandoned
	Internet network issue	Test internet connection prior to use	Call attempt abandoned	Install locate and recovery procedures
The softphone is operational less than 95% of the time	Software not updated	Assure software upgrades installed	Operation fails	Install supplemental support procedures
	Device is faulty	Test device prior to operation	Operation proceeds intermittently	

14.4 Application of the Human Views to Phishing Risk

Humans are considered the weakest link in a phishing attack (Boulton 2017). Phishing's main objective is to deceive individuals into giving up sensitive information (Xu and Zhang 2012). By clicking on a deceptive link in an email, or opening a document or a file that is attached to the email, a human user can infect a computer and connected systems almost immediately in some cases. In other instances, the fraudulent link will ask them to log into a familiar site using their password, except it is a false site that captures their personnel information, credit card and/or banking information. If the user responds to the email, the individual may divulge personnel or organizational sensitive information to the hackers.

Hence, it is paramount in managing phishing risk to include the human interactions with the phishing process. Figure 14.3 is a Human Network model describing the interactions of humans with the phishing process from the time the user first clicks on a phishing email to when the individual becomes a victim, or "hooked and caught". The diagram illustrates the multiple opportunities the user has to prevent becoming a phishing victim. The Human Network sequence diagram can identify the points where risk mitigations can be applied to prevent a successful phishing attack.

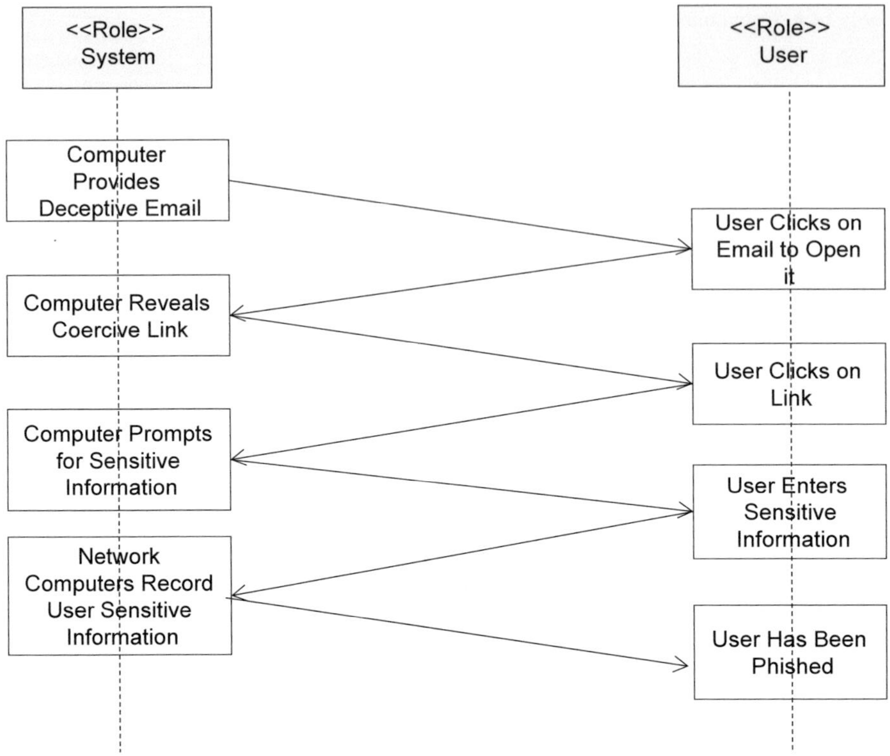

Fig. 14.3 Sequence diagram of a phishing event

There are typically three components that most phishing attacks rely upon to successfully achieve its objectives: the lure, hook, and catch (Chaudhry et al. 2016). The lure is anything that attracts the attention of the individual; it is most often an email that offers a great reward or describes a potential issue for a financial account. The lure stresses the need for urgency and encourages the individual to act expediently so as not to miss the offer or to avoid negative consequences. The hook occurs the moment the individual takes action. Once the link is clicked with the intention of completing the action, the individual has been hooked. The hook is the reaction to lure; if the individual responds to the link, it a positive reaction for the hacker, but if the individual declines, it is a positive reaction for the organization. Finally, the catch is the information that an individual divulges to the hacker. This may include a range of personal or organization information that can be used for identity theft or to access other organizational systems.

Figure 14.4 shows a Task model of the activities an individual performs while interacting with the phishing system. It indicates the role of the individual as a "gatekeeper" of both personal and organizational information, as well as the role of the attacker as a "thief" to obtain sensitive information. Importantly, the Task

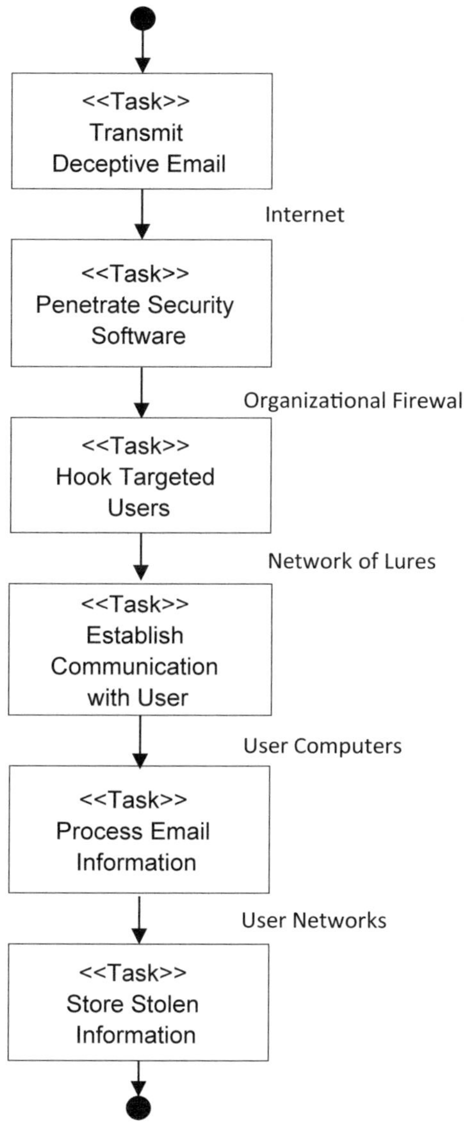

Fig. 14.4 Activity diagram to steal sensitive information

model includes the system interfaces the individual is interacting with when the phishing attack occurs, identifying these system interfaces as possible points of risk mitigation strategies. The strategies may include training the individual on how to more accurately identify emails and phone calls that are not legitimate and possibly attempts to phish sensitive information.

14.5 Summary

The focus of a system architecture development is to address stakeholder concerns, perform trade-off analysis among competing priorities, and provide the baseline for further system development. Risk management augments this process by trying to address various failure scenarios based on stated and implied goals that the system may encounter during its use. The Human Viewpoint methodology can be used to capture human focused data and analyses for socio-technical systems; the Human View models can be expanded to include details that support risk management activities focused on human operators. The seven guiding questions for operational risk management have been aligned with the Human Viewpoint methodology and can be used to identify and mitigate the human role in various risk scenarios.

References

Basel II (2004) International convergence of capital measurement and capital standards: a revised framework. Bank for International Settlements Press & Communications CH-4002, Basel, Switzerland

Boulton C (2017) Humans are (still) the weakest cybersecurity link. https://www.cio.com/article/3191088/security/humans-are-still-the-weakest-cybersecurity-link.html. Accessed 16 Oct 2018

Chaudhry J, Chaudhry S, Rittenhouse R (2016) Phishing attacks and defenses. Intl J Secur Appl 10(1):247–256

Garvey P (2008) Analytical methods for risk management: a systems engineering perspective. Chapman-Hall/CRC-Press, Boca Raton, FL

Handley H, Knapp B (2014) Where are the people? The human view approach for architecting and acquisition. Def Acquis Res J 21(4):851–874

Handley H, Kandemir C, Dario E (2015) Warrior information network-tactical human view models & analysis. Technical report, Old Dominion University, Norfolk, VA

Kaplan S (1997) The words of risk analysis. Risk Anal 17(4)

Pinto A, Tolk A, Landaeta R (2010) Goal approach to risk scenario identification in systems development. In: Proceedings of the 31st national conference of the American Society for Engineering Management, Arkansas, USA, 13–16 October 2010

Xu Z, Zhang W (2012) Victimized by phishing: a heuristic-systematic perspective. J Internet Bank Commerce 17(3):1–15

Chapter 15
Conclusion: The Human Viewpoint Methodology for Socio-technical Systems

Abstract The final chapter in the book reiterates the need for the Human Viewpoint methodology to collect human focused data, complete pertinent analyses, and provide Fit for Purpose models that identify the human considerations of the socio-technical system. It identifies the contributions of the Human Views to both system architecting and to Human System Integration. It provides insights to the use of the Human Views in achieving executable architectures, i.e., simulations derived directly from the static views. Finally, the chapter summarizes the main points of the book.

Keywords System architectures · Human system integration · Executable architectures

15.1 Introduction

Human System Engineering focuses on including human capabilities and constraints within the system engineering process in order to improve the performance of socio-technical systems. Traditionally system design focused on the functional and technical aspects of the system, postponing human operator concerns until the later in the system development. The Human Viewpoint provides a methodology to integrate human considerations into the system architecting phase, early in the system engineering process. The resulting Human Views capture the capabilities and limitations of the human operators interacting with the system. The Human Viewpoint methodology facilitates the collection of human focused data, completing focused analyses, and providing Fit for Purpose models that identify the human considerations of socio-technical systems.

15.2 Contribution to System Architectures

System architecting results in a system description that is communicated through a series of different perspectives or viewpoints. The architecture description provides the realization of an operational concept as a baseline system for discussion,

H. A. H. Handley, *The Human Viewpoint for System Architectures*, Topics in Safety, Risk, Reliability and Quality 35, https://doi.org/10.1007/978-3-030-11629-3_15

decision-making, and further development. Traditionally the human component interacting with the system was not explicitly included in this process. While there are opportunities to capture some human parameters within other architecture views, there was not a dedicated viewpoint to address stakeholder concerns focused on the operators of the system. By including a separate Human Viewpoint, the architecture description is augmented with the relevant human focused data necessary to have a complete representation of the socio-technical system.

The Human Viewpoint is an integrated architecture, i.e., there are associations from the elements captured in the Human Views to elements captured in the other architecture viewpoints. This provides linkages between the mission capabilities, operational activities, and system functions to the operator roles and tasks. Changes made in the other viewpoints may induce changes to the human roles, responsibilities, and tasking, as well as implications for training, personnel selection, and crew assignments. The integration of the Human Views with the other architecture views allows human concerns to be identified and addressed during system development.

The Human Views are considered Fit for Purpose views. Fit for Purpose views were introduced into architecture descriptions to allow the architecture to focus on collecting data and creating views that responded to stakeholder concerns. Fit for Purpose views do not necessarily align with the standard model templates, and the rendering of the views may result in unique representations of the architecture data. The Fit for Purpose approach emphasizes formatting architectural data to support analysis and decision-making and presenting the results in a meaningful and useful way. While the use of Fit for Purpose views is encouraged, there is very little guidance on methods to create the views. The Human Viewpoint methodology was developed specifically to create Fit for Purpose Human Views and can be applied across other viewpoints. The Human Viewpoint methodology encourages architecture development to focus on collecting data and creating views that are necessary for the stakeholder requirements, and creating unique presentations to support decision-making rather than relying on legacy model templates. Fit for Purpose views can be created as needed to ensure that the architectural data is easily understood.

The Human View methodology focuses on collecting and organizing human focused data, identifying the important relationships between the data elements, and rendering views of the data to provide models of the human system. The models are rendering using SysML templates so that the visual modeling representations are similar to other architecture products. This information can then be used for analysis based on different scenarios or use cases. The relationships identified in the Human View analysis can be used to vary conditions such as role to task assignment to evaluate the impact of human constraints and limitations on overall system performance.

15.3 Contribution to Human System Integration

The Human Viewpoint facilitates including Human System Integration (HSI) into the system architecting process by promoting early consideration of human issues. The Human Views provide role and task definitions and system interface requirements. The Human Viewpoint can reduce system risk due to technical design problems by communicating information about the needs and constraints of the human operator or user. HSI issues due to reduced crew sizes, dependence on contractor support, and increased training requirements result in increased performance times, error rates, and high crew workloads. Using the Human Viewpoint, many of these issues can be investigated while system design decisions can still be influenced in order to save on system redesign costs and initial poor system performance. The Human Views support personnel planning for the deployed system by providing an early assessment of the task allocation, role requirements and essential training.

Additionally, initial trade-off analyses between the different HSI domains can be completed using the Human Viewpoint. Manpower assessments are driven by the number of people required by the system; increasing the functionality of the system, which increases the number of tasks and roles, will increase manpower requirements. Personnel capabilities focuses on the types of people needed to operate the system by detailing the knowledge, skills and abilities (KSA) required for the defined roles. As personnel are asked to do more diverse tasks, there may be an increase in the KSA requirements to fill each role; this may require changes in the personnel types identified for the system. Training evaluates the gaps between the skills that personnel already have and the requirements for the new system. This information may place additional limitations on role to task assignments.

It is not necessary to complete the full set of Human Views to benefit from a human focused architecting effort. Each individual Human View captures a "snapshot" of different aspects of the socio-technical system and can add value to the architecture description and support HSI focused analyses. For example, the Tasks view captures the human activities of a system. These tasks can be described in terms of a sequence diagram providing a time-based ordering of the tasks. This offers an indication of how a given sequence of tasks will perform and the performance predictions for alternative sequences of tasks can be compared. An analyses of the Roles view may result in recommendations to reallocate tasks to other roles based on workload, skill requirements, or locations. For network based systems, an analysis of the Human Network may result in coordination requirements for distributed team members to help define responsibilities and information sharing. Even using a subset of the Human Viewpoint models provides the opportunity to capture and organize diverse human information to assess the evolving design and recommend improvements to the socio-technical system.

15.4 Advances Towards Executable Architectures

The Human Views Dynamics provides a simulation model that can be used to predict operator performance; the simulation model is designed based on the information provided in the static views. Executable architectures are techniques to extend existing architecture frameworks to allow dynamic analysis of system behavior within the architecting environment. The Human Views take the first step towards this capability by rendering the models in the SysML language. Because SysML is based on an underlying formalism, there is some ability to semi-automate the generation of executable models from the data in the Human Views, insuring consistency between the static and dynamic models.

Executable architectures provide validation, i.e., correctness of the data, and verification, i.e., uniformity in the use of the data, to check for logical consistency within the set of models. An executable architecture allows the analysis of the dynamic behavior of the modeled system to identify errors not easily found in the static views. The automatic generation of an executable model within the architecture context, rather than exporting to a simulation environment, is necessary to maintain the completeness and conformance of the architecture data. When a simulation is created by extracting data directly from the Human Viewpoint database, the Human Views provide the data to model the logical behavior of the socio-technical system.

15.5 Summary

The Human Views were developed as a way to collect and categorize human focused information for use in system architecture developments. The Human Viewpoint methodology can provide the data and analyses required to evaluate the impact of changes to the socio-technical system. The Human Views are integrated with the rest of the architecture views and can provide tradeoff analyses between competing demands. The Human Viewpoint models help decision makers understand the key human components of a system by providing a representation of roles, tasks, work processes as well as limitations and constraints. The Fit for Purpose customizable views are created to address specific stakeholder questions and display specific sets of data; these models can then be used to create a description of the baseline "as-is" socio-technical system and used to evaluated potential "to-be" system implementations.

The Human Viewpoint methodology supports a Model Based System Engineering (MBSE) approach. It focuses on collecting and organizing human focused data, identifying the important relationships between the data elements, and rendering views of the data to provide models of the socio-technical system. These models can be rendered using visual representations similar to other architecture products. The System Modeling Language (SysML) is a general way to represent system constructs using visual models which can be adapted to include human focused

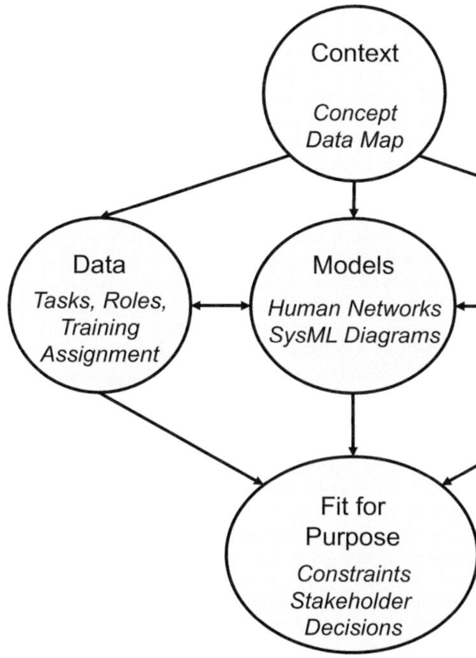

Fig. 15.1 The formalized Human Viewpoint methodology

data. Using SysML models offers a common language for both the technical factors as well as the human concerns for the developing system. A system architecture description captured in a cogent set of models provides inherent rigor by documenting the architecture with a set of standard representations.

The initial proposal for the Human Views to augment system architectures suffered from the lack of a formal implementation methodology and inconsistent views. The Human Viewpoint methodology described in this book addresses these concerns and provides a rigorous process to guide the socio-technical architecture development. The Human Viewpoint addresses stakeholder concerns to help evaluate tradeoffs between the competing goals of the human operator and the system functionality for the socio-technical system. It provides a data repository of human focused data that can be used at multiple levels of concern, aligned with domain specific frameworks, and indicates design completeness and compliance as well as evaluation of human performance. By contributing to an integrated set of architecture models, the Human Views support design decisions for the socio-technical system. Additionally, the Human Viewpoint provides a self-contained set of data and relationships that allows decoupling the human sub system for analysis as well as integration within the larger system for a total system of systems approach. The formalized Human Viewpoint methodology is shown in Fig. 15.1.

Appendix A
Human View Data for the Commander's Daily Update Brief Example

A.1 Introduction

The original Human Viewpoint described a set of outcome products similar to other architecture viewpoints at the time. These products were diagrams, tables, figures or other descriptions of human focused data. In the context of today's Human Viewpoint, these products are equivalent to the data tables used to collect the human focused data that can then be rendered in different models, providing the individual Human Views. This appendix presents a comprehensive example of the human view data from the Commander's Daily Update Brief Process (NATO 2010).

A.2 Example

The Commander's Daily Update Brief is an operational brief that provides updates regarding the readiness and operational assets throughout the command, with a focus on the previous 24 hours and the next 24 hours. A Commander's Daily Update Brief Process is in place in virtually every US military command. The staff process that produces the brief includes analyzing data sources, creating Microsoft Power Point slides, and numerous review cycles. Historically, the production of the brief has been a manual, staff intensive process that often resulted in static information which was often several hours old. Prior to the implementation of the Integrated Interactive Data Briefing Tool (IIDBT), this process consumed staff members working the night shift, while the day shift's personnel devoted the morning hours to its production (Pester-DeWan et al. 2003). The IIDBT automated the data gathering process using Web services that pull data directly from

H. A. H. Handley, *The Human Viewpoint for System Architectures*, Topics in Safety, Risk, Reliability and Quality 35, https://doi.org/10.1007/978-3-030-11629-3

authoritative sources; the automation of these formerly manual processes saved the staff an estimated 3.5 h a day while at the same time allowing them to present more current information (Higgins and Hall 2004). While production time was cut significantly, the process is still largely stove-piped along functional area divisions. Coalescing the information for the brief typically requires 15–20 people and numerous reviewers from various functional areas to create a series of Power Point slides that are organized into a single presentation that is catered to the commander's information requirements (Handley and Heacox 2005).

A.3 Concept

The Concept provides the different conditions under which the process will be examined and defines the outcomes to be measured. For this example, the area of concern is the performance of the briefing process with and without the IIDBT under different conditions of operational tempo, i.e., the rate at which information is available, and levels of connectivity, i.e., the number of communication channels available for information exchange. The Concept data is shown in Table A.1.

A.4 Tasks

The Tasks data decomposes higher level activities into discrete tasks that can be assigned to roles. In Table A.2, the top row is the high-level activities, and each associated column is the decomposed set of tasks required to accomplish the activity.

Additionally, a second set of data was collected that identifies the system interfaces that are required by the tasks. Table A.3 identifies the systems required by the high-level activities. This is important for the stakeholder evaluation of the IIDBT system.

Table A.1 Concept data for Commander's Daily Update Brief

Process with IIDBT		Process without IIDBT		Performance measure
Operational tempo	Connectivity	Operational tempo	Connectivity	Brief completion time
Low	Low	Low	Low	
Low	High	Low	High	
High	Low	High	Low	
High	High	High	High	

Table A.2 Task decomposition for Commander's Daily Update Brief

1.0 Identify new information for assigned topics	2.0 Create assigned slides	3.0 Approve slides at cell level	4.0 Compile the briefing form posted slides	5.0 Approve slides at command level	6.0 Brief commander and staff
Select topics for briefing content	Obtain templates for briefing	Advise reviewers of readiness	Access slides posted by assigned cells	Advise reviewers of readiness	Send link for collaborative session
Review previously submitted data	Import data	Review slides	Assess if all slides have been posted	Review slides	Access session
Identify data sources for relevant updates	Create slide	Provide updates and comments	Notify appropriate cell staff that slides are due	Provide updates and comments	Initiate collaborative session
Access sources and identify information	Revise slides and notes	Review comments	Access status of requested slides	Review comments	Take roll call
	Assess currency of information	Assess need for more info	Notify BWC to proceed without slides	Access and revise slides	Present the brief
	Assess accuracy of fields and spelling	Access sources and identify new information	Arrange posted slides in order for briefing	Post reviewed slides	Discuss issues and implications
	Revise slide fields and spelling	Import data		Ensure order and content of posted slides	Determine action items
	Assess need to make changes to notes	Assess need to make changes to slides			Distribute action items
	Revise slide notes	Access and revise slides			
	Assess need for sharing with foreign partners	Post reviewed slides			
	Assess compliance with disclosure policies				
	Post completed slide				

Table A.3 System interface matrix for Commander's Daily Update Brief

Tasks	System							
	Crisis action page	Digital ROE	SIPRNET	Electronic bookmarks	IIDBT	Shared folder	Email	Same time
1.1 Select topics for briefing content	x	x						
1.2 Review previously submitted data	x							
1.3 Identify data sources for relevant updates			x					
1.4 Access sources and identify information			x	x				
2.1 Obtain templates for briefing					x			
2.2 Import data			x					
2.3 Create slide					x			
2.4 Revise slides and notes					x			
2.7 Revise slide fields and spelling					x			
2.9 Revise slide notes					x			
2.11 Assess compliance of data with disclosure policies							x	
2.12 Post completed slide						x		
3.1 Advise reviewers of readiness							x	
3.2 Review slides						x		
3.3 Provide updates and comments							x	
3.4 Review comments							x	
3.6 Access sources and identify new information			x	x				
3.7 Import data				x				
3.9 Access and revise slides					x			

(continued)

Table A.3 (continued)

Tasks	System							
	Crisis action page	Digital ROE	SIPRNET	Electronic bookmarks	IIDBT	Shared folder	Email	Same time
3.10 Post reviewed slides	x							
4.1 Access slides posted by assigned cells	x							
4.3 Notify appropriate cell staff that slides are due							x	
4.5 Notify BWC to proceed without slides							x	
4.6 Arrange posted slides in order for briefing					x			
5.1 Advise reviewers of readiness							x	
5.2 Review slides	x							
5.3 Provide updates and comments							x	
5.4 Review comments							x	
5.5 Access and revise slides					x			
5.6 Post reviewed slides	x							
5.7 Ensure order and content of posted slides					x			
6.2 Access Session								x
6.3 Initiate Collaborative session								x
6.4 Take roll call								x
6.5 Present the brief	x							
6.6 Discuss issues and implications								x
6.7 Determine action items								x
6.8 Distribute action items							x	

A.5 Roles

The Roles table defines the roles for the Commander's Daily Update Brief process. Table A.4 lists the roles required by this process with some associated attributes such as multiplicity, competency and authority.

The role to task assignment matrix allocates task responsibilities to the different roles, as shown in Table A.5. This table indicates the role responsibilities for the process tasks. Note that some tasks are assigned to role "teams" while others are assigned to individual roles.

A.6 Training

For the Commander's Daily Update Brief example, the Training data focuses on the required qualifications for personnel to assume the defined roles. These qualifications, which include rank, military designator, clearance level and location, are shown in Table A.6.

A.7 Human Network

The Human Network data focuses on the information exchange requirements between the roles to support task completion. Roles that need to exchange information, along with the systems that are used, are shown in Table A.7.

A.8 Metrics

Human performance objectives, indicators and risks associated with specific tasks are identified in Table A.8. The Metrics data are used in the evaluation of alternative instantiations of the process under the different technical conditions of the concept.

A.9 Summary

This example presents the different sets of data that can be collected to support a Human Viewpoint development. In the original Human Viewpoint description, these tables would have been considered outcome products. With the evolution of the Human Viewpoint, these tables now provide the data to render relevant models to support the Human Viewpoint analysis.

Table A.4 Roles for Commander's Daily Update Brief

Title	Multiplicity	Team	Competency	Authority
Commander	Individual		GW36—guiding directing, and motivating subordinates	Level 0
Director of manpower and personnel	Individual	Cell directors	GW36—guiding directing, and motivating subordinates	Level 1
Update development staff	Group	CFMCC staff	GW09—analyzing data or information	Level 2
Director of intelligence	Individual	Cell directors	GW36—guiding directing, and motivating subordinates	Level 1
Update development staff	Group	CFMCC staff	GW09—analyzing data or information	Level 2
Special security officer	Individual		GW07—evaluating information to determine compliance with standards	Level 2
Director of operations	Individual	Cell directors	GW33—coordinating the work and activities of others	Level 1
Update development staff	Group	CFMCC staff	GW09—analyzing data or information	Level 2
Current operations (COPS)	Individual		GW26—communicating with supervisors, peers, or subordinates	Level 2
Battle watch captain (BWC)	Individual		GW26—communicating with supervisors, peers, or subordinates	Level 2
Director of logistics	Individual	Cell directors	GW36—guiding directing, and motivating subordinates	Level 1
Update development staff	Group	CFMCC staff	GW09—analyzing data or information	Level 2
Director of planning	Individual	Cell directors	GW36—guiding directing, and motivating subordinates	Level 1
Update development staff	Group	CFMCC staff	GW09—analyzing data or information	Level 2
Director of C4I	Individual	Cell directors	GW36—guiding directing, and motivating subordinates	Level 1
Update development staff	Group	CFMCC staff	GW09—analyzing data or information	Level 2
Director of training	Individual	Cell directors	GW36—guiding directing, and motivating subordinates	Level 1
Update development staff	Group	CFMCC staff	GW09—analyzing data or information	Level 2
Director of experimentation	Individual	Cell directors	GW36—guiding directing, and motivating subordinates	Level 1
Update development staff	Group	CFMCC staff	GW09—analyzing data or information	Level 2

Table A.5 Task responsibility matrix for Commander's Daily Update Brief

Tasks	Responsibility							
	Director operations	CFMCC staff	Cell directors	Special security officer	Battle watch captain	COPS	Remote staff	Commander
1.1 Select topics for briefing content	x							
1.2 Review previously submitted data		x						
1.3 Identify data sources for relevant updates		x						
1.4 Access sources and identify information		x						
2.1 Obtain templates for briefing		x						
2.2 Import data		x						
2.3 Create slide		x						
2.4 Revise slides and notes		x						
2.5 Assess currency of information		x						
2.6 Assess accuracy of fields and spelling		x						
2.7 Revise slide fields and spelling		x						
2.8 Assess need to make changes to notes		x						
2.9 Revise slide notes		x						
2.10 Assess need for sharing with foreign partners		x						
2.11 Assess compliance of data with disclosure policies				x				
2.12 Post completed slide		x						
3.1 Advise reviewers of readiness		x						
3.2 Review slides			x					
3.3 Provide updates and comments			x					

(continued)

Table A.5 (continued)

Tasks	Responsibility							
	Director operations	CFMCC staff	Cell directors	Special security officer	Battle watch captain	COPS	Remote staff	Commander
3.4 Review comments		x						
3.5 Assess need for more info		x						
3.6 Access sources and identify new information		x						
3.7 Import data		x						
3.8 Assess need to make changes to slides		x						
3.9 Access and revise slides		x						
3.10 Post reviewed slides		x						
4.1 Access slides posted by assigned cells					x			
4.2 Assess if all slides have been posted					x			
4.3 Notify appropriate cell staff that slides are due					x			
4.4 Access status of requested slides		x						
4.5 Notify BWC to proceed without slides		x						
4.6 Arrange posted slides in order for briefing					x			
5.1 Advise reviewers of readiness					x			
5.2 Review slides			x					
5.3 Provide updates and comments			x					
5.4 Review comments		x						
5.5 Access and revise slides		x						

(continued)

Table A.5 (continued)

Tasks	Responsibility							
	Director operations	CFMCC staff	Cell directors	Special security officer	Battle watch captain	COPS	Remote staff	Commander
5.6 Post reviewed slides		x						
5.7 Ensure order and content of posted slides					x			
6.1 Send link for collaborative session						x		
6.2 Access session	x		x		x		x	x
6.3 Initiate collaborative session						x		
6.4 Take roll call					x			
6.5 Present the brief	x							
6.6 Discuss issues and implications								x
6.7 Determine action items								x
6.8 Distribute action items					x			

Table A.6 Role requirements for Commander's Daily Update Brief

Code	Title	Rank	Designator	Clearance	Location
J00	Commander (CDR)	O-9	1310	TS	Afloat
J1	Director of manpower and personnel	O-5	1315	S	Ashore
J2	Director of intelligence	O-6	1630	TS/SCI	Ashore
SSO	Special security officer	E-6	CTA	TS/SCI	Ashore
J3	Director of operations (OPS)	O-6	1110	TS/SCI	Afloat
J33	Current operations (COPS)	O-5	1147	TS	Afloat
BWC	Battle watch captain	O-4	1320	TS	Afloat
J4	Director of logistics	O-6	3100	S	Ashore
J5	Director of planning	O-5	1310	TS	Afloat
J6	Director of C4I	O-6	1120	TS	Ashore
J7	Director of training	O-5	1310	S	Ashore
J9	Director of experimentation	O-6	1320	TS	Ashore

Table A.7 Information coordination for Commander's Daily Update Brief

Operations/Tasks in this activity	Technology/Applications	Information flow	
		From	To
1.1 Select topics for briefing content	Routine: CAS Web (Battle Rhythm) Special: e-mail, chat	OPS	CFMCC Staff
1.2 Review previously submitted data	CAS Web (view previous posting), e-mail, chat	(CFMCC Staff work in progress)	
1.3 Identify data sources for relevant updates	DISA Federated or other search tool, electronic bookmarks, e-mail, chat	(CFMCC Staff work in progress)	
1.4 Access sources and identify information	DISA Federated tool, SIPRNET	(CFMCC Staff work in progress)	
2.1–2.3 Import data into templates		(CFMCC Staff work in progress)	
2.4–2.9 Ensure accuracy of information and presentation, and post updated info	IIDBT, PowerPoint, Cell's private shared folders, Public shared folders	(CFMCC Staff work in progress)	
2.10 Assess need for sharing with foreign partners	e-mail, chat	CFMCC Staff	SSO
2.11 Assess compliance of data with disclosure policies	Public shared folders, e-mail chat	SSO	CFMCC Staff
2.12 Post completed slide	DISA Federated or other search tool, electronic bookmarks, e-mail, chat, Public shared folders, SIPRNET, IIDBT, Cell's private shared folders	(CFMCC Staff work in progress)	
3.1 Advise reviewers of readiness	e-mail, chat	CFMCC Staff	Cell Director
3.2–3.3 Conduct cell-level review and provide feedback	Public shared folders, e-mail, chat	Cell Director	CFMCC Staff
3.4–3.10 Review feedback, revise materials per review as necessary and re-post	DISA Federated or other search tool, electronic bookmarks, e-mail, chat, Public shared folders, SIPRNET, IIDBT, Cell's private shared folders	(CFMCC Staff work in progress)	
4.1–4.2 Assess if required material has been posted	Public shared folders	(BWC work in progress)	
4.3 Request additional materials	e-mail, chat	BWC	CFMCC Staff
4.4–4.5 Develop additional material per request, re-post and notify	DISA Federated or other search tool, electronic bookmarks, e-mail, chat, Cell's private shared folders, Public shared folders, SIPRNET, IIDBT	CFMCC Staff	BWC

(continued)

Table A.7 (continued)

Operations/Tasks in this activity	Technology/Applications	Information flow	
		From	To
4.6 Arrange posted slides in order for briefing and notify reviewers	Public shared folders, e-mail, chat	BWC	Cell Directors
5.1–5.3 Conduct command-level review and provide feedback	Public shared folders, e-mail, chat	Cell Directors	CFMCC staff, (copy BWC)
5.4–5.5 Review feedback, revise materials per review as necessary, re-post and notify	DISA Federated or other search tool, electronic bookmarks, e-mail, chat, Cell's private shared folders, Public shared folders, SIPRNET, IIDBT	CFMCC Staff	BWC
5.6–5.7 Finalize and post material for briefing	Public shared folders, CAS Web, e-mail, chat	BWC	COPS
6.1 Send link for collaborative session	e-mail, chat	COPS	OPS, Cells, CDR, Remotes, BWC
6.2 Access session	VOIP/VTC, CAS Web, Same-Time, IWS, e-mail, chat	OPS, Cells, CDR, Remotes, BWC	COPS
6.3 Initiate collaborative session	VOIP/VTC, Same-Time, IWS	COPS	OPS, Cells, CDR, Remotes, BWC
6.4 Take roll call	VOIP/VTC, Same-Time, IWS	BWC	OPS, Cells, CDR, Remotes, COPS
6.6–6.7 Present the brief, discuss implications	VOIP/VTC, CAS Web, Same-Time, IWS	OPS, Cells, Remotes	CDR
6.7 Determine COA	VOIP/VTC, CAS Web, Same-Time, IWS	CDR	OPS, Cells, Remotes, BWC
6.8 Distribute decision/action items	e-mail, chat	BWC	OPS, Cells, Remotes, CDR

Table A.8 Metrics for Commander's Daily Update Brief

Tasks	Objectives	Indicators	Risks
1.0 Identify new information for assigned topics	1. Relevant new information is identified; 2. Requests for briefing, both standard and special are acted upon		
1.1 Select topics for briefing content		Brief development is started within time targets	Missed trigger to begin process
1.3 Identify data sources for relevant updates		Information identified is the most up-to-date available	Topical requirements are misunderstood
1.4 Access sources and identify information		Information identified is relevant to the situation	Data sources are not accessible
2.0 Create assigned slides	1. All available information required to respond to situation is included; 2. Preparation is within time limits		
2.2 Import data			There is a lack of connectivity to sources
2.3 Create slide		Information on the slide is relevant to the situation	Data updates are not imported
2.5 Assess currency of information		Information on the slide is the most up-to-date available	
2.6 Assess accuracy of fields and spelling			The request for a special format is missed
2.12 Post completed slide		Slide preparation is within time limits	Development schedule is not followed
3.0 Approve slides at cell level	1. Adherence to the development schedule is maintained; 2. The review process results in higher quality slides		
3.1 Advise reviewers of readiness		Reviewers are available when needed	Reviewers are not available
3.2 Review slides			Review is a technicality

(continued)

Table A.8 (continued)

Tasks	Objectives	Indicators	Risks
		Requested slides are posted and accessible	
3.3 Provide updates and comments		Accuracy of information is improved	
3.6 Access sources and identify new information		Information identified is relevant to the situation	Data sources are not accessible
3.10 Post reviewed slides		Slides are reviewed and changed within time targets	
4.0 Compile the briefing form posted slides	1. Adherence to the development schedule is maintained; 2. The compiled brief contains all requested slides		
4.1 Access slides posted by assigned cells		Requested slides are posted and accessible	Posted slides are inaccessible/ incompatible
4.3 Notify appropriate cell staff that slides are due			Missed trigger to begin brief development process
4.4 Access status of requested slides		Requested slides are posted and accessible	
4.6 Arrange posted slides in order for briefing		Brief is completed within time targets	
5.0 Approve slides at command level	1. Adherence to the development schedule is maintained; 2. The review process results in higher quality slides		
5.1 Advise reviewers of readiness		Reviewers are available when needed	Reviewers are not available
5.2 Review slides		Requested slides are posted and accessible	Review is a technicality
5.3 Provide updates and comments		Accuracy of information is improved	

(continued)

Table A.8 (continued)

Tasks	Objectives	Indicators	Risks
5.6 Post reviewed slides		Slides are reviewed and changed within time targets	
5.7 Ensure order and content of posted slides		Brief is compiled within time targets	
6.0 Brief commander and staff	1. Briefing schedule is maintained; 2. Commander gains up to date SA of the situation; 3. Follow-on tasks are assigned		
6.1 Send Link for collaborative session		The brief is conducted within the time target	Delays cause the briefing to be late
6.2 Access Session		All staff are able to access the session	Staff are unable to access the brief
6.4 Take roll call		All requested staff are present	
6.5 Present the brief		Current information is presented	The most current information is not presented
6.6 Discuss issues and implications		Relevant information is presented	
6.7 Determine action items		Action items are developed	
6.8 Distribute action items			Action items are not relayed

References

Handley H, Heacox N (2005) DoDAF products for the commander's daily update brief process, baseline version. Technical Report, Pacific Science & Engineering, San Diego, CA

NATO (2010) Human systems integration for network centric warfare. Technical Report, North Atlantic Treaty Organization Research Technology Organization, TR-HFM-155, Brussels, Belgium

Higgins E, Hall J (2004) The IIDBT meets the demands of modern warfare with speed and accuracy. Chips Magazine July–September 2004

Pester-DeWan J, Moore R, Morrison J (2003) Knowledge engineering for command and control transformation. Technical Report, Pacific Science & Engineering, San Diego, CA

Appendix B
Human View SysML Models for React to Contact Example

B.1 Introduction

A sample scenario, React to Contact, was used to create a demonstration set of Human View models using System Modeling Language (SysML) diagrams. This effort specifically focused on attempting to incorporate all types of SysML diagrams in order to evaluate their utility for use as alternative template options for the Human View models.

B.2 React to Contact Scenario

The React to Contact scenario is an action sequence that an Army Platoon takes to return fire while seeking cover and concealment. The squads are moved by their leaders to establish firing positions to suppress the enemy. This is done by identifying and assuming a firing position that is conductive to achieving fire superiority (Taylor 2014). The Package Diagram, shown in Fig. B.1, indicates the organization of the Human View models for this scenario.

The stakeholder interest in the React to Contact scenario is understanding the cognitive and physical aspects of the soldier actions with respect to different rifle types (Taylor 2014). Table B.1 identifies the different types of physical and cognitive actions under consideration (DOA 2007).

H. A. H. Handley, *The Human Viewpoint for System Architectures*, Topics in Safety, Risk, Reliability and Quality 35, https://doi.org/10.1007/978-3-030-11629-3

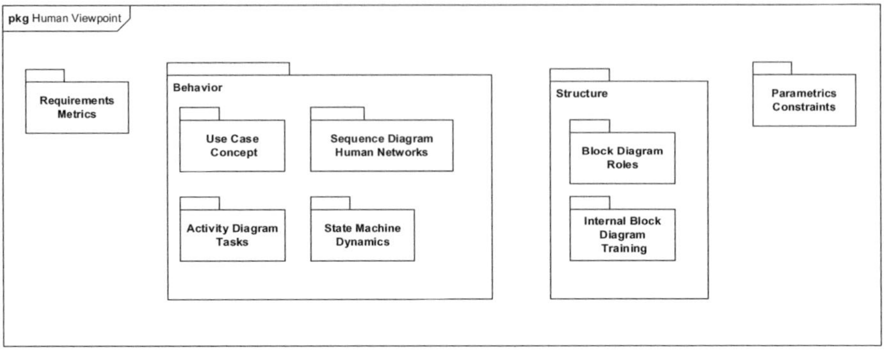

Fig. B.1 Package diagram for "React to Contact" Human Viewpoint

Table B.1 Rifle physical and cognitive actions

	Physical actions	Cognitive actions
1. Detect targets depending on position, skill in scanning, and ability to observe the area and recognize target indicators		X
2. Select position; the position should protect from enemy fire and observation, yet allow effective fire on targets in the sector of fire		X
3. Determine range to targets		X
4. Identify targets in your designated sector of fire		X
5. Fire on targets using correct fundamentals of marksmanship and appropriate aiming and engagement techniques	X	

B.3 Concept

The Concept view is visualized using a SysML Use Case diagram. It represents how the React to Contact scenario interacts with external entities, i.e., the soldiers. Use Cases are behavior diagrams and represent the highest level of abstraction of the system. Figure B.2 shows the Use Case for the Human View Concept. On the left-hand side is the Use Case with the React to Contact system boundary and the Infantry Platoon as the actor that interacts with the system. On the right-hand side is a decomposition of the first series in the React to Contact process, with the individual tasks that will occur and the roles that will interact with these tasks.

B.4 Tasks

SysML Activity diagrams represent behavior in terms of the ordering of actions based on the availability of inputs, outputs, and controls, and how the actions transform the inputs to outputs. For the Tasks view, an Activity diagram is used to

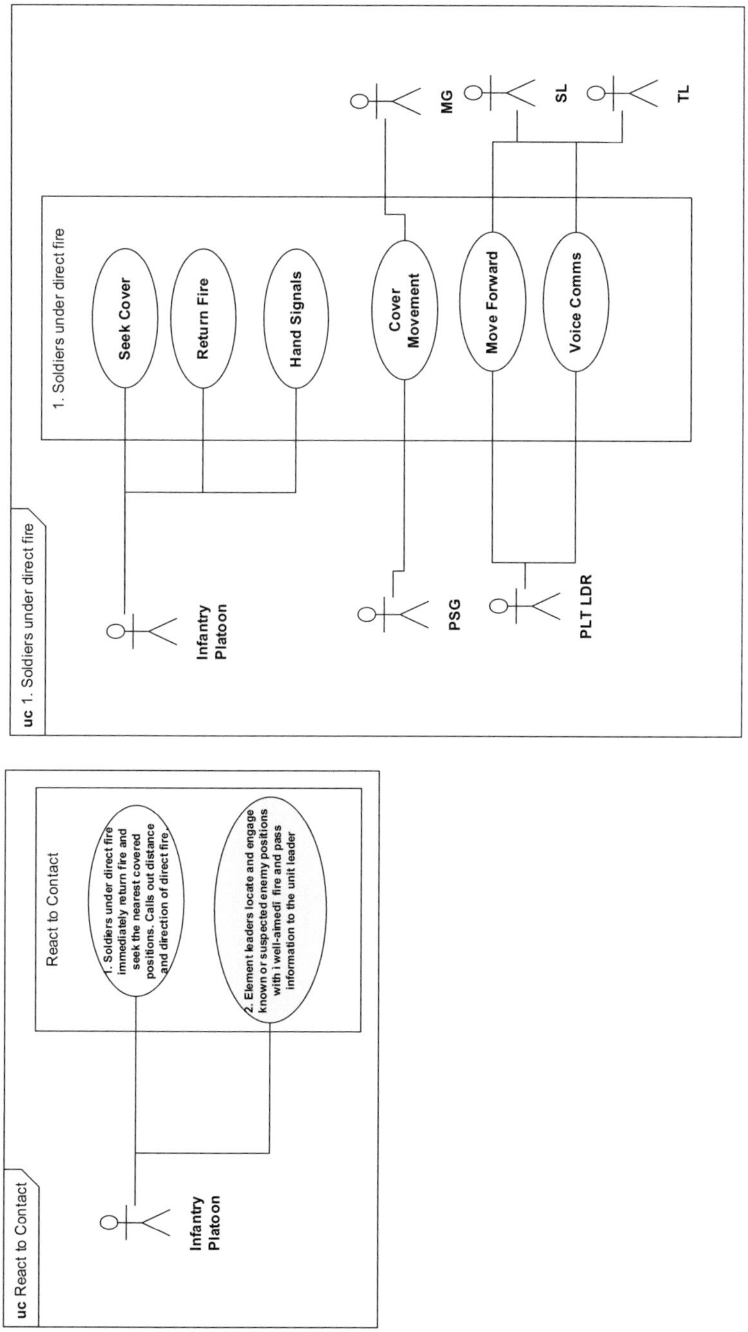

Fig. B.2 Use case diagram for Human View concept

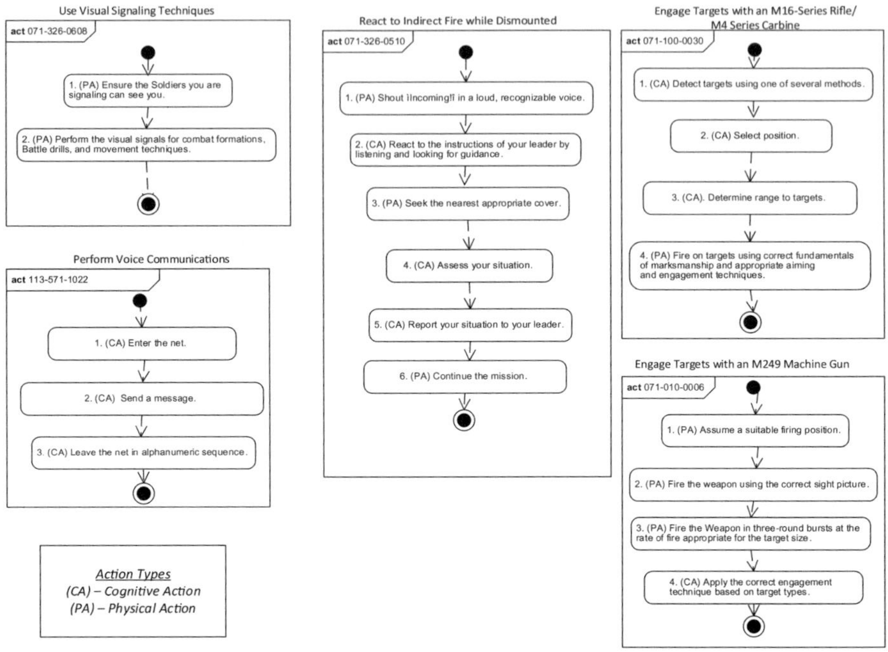

Fig. B.3 Activity diagram for Human View tasks

detail the different task processes in the React to Contact scenario. Figure B.3 shows the Activity diagrams for "Soldiers under Direct Fire". For each task, a separate Activity diagram is included, with the task number, and the actions to be performed labeled with either Physical Action (PA) or Cognitive Action (CA). Also note that the name of the task is given above the diagram, while its reference number is used in the title of the diagram.

B.5 Roles

For the Roles view, a SysML Block Definition diagram was used. Block diagrams represents structural elements and can include information on their composition and classification, as well as associations. Block diagrams provide a general-purpose capability to model system components. Figure B.4 uses a Block diagram to show the relationship of the roles within the Platoon. The Platoon leader commands two squads, each of which consists of two teams. Additionally, the Platoon leader has a staff of personnel that report directly. The personnel qualifiers are provided below each role abbreviation and the legend on the side of the diagram provides the full role names.

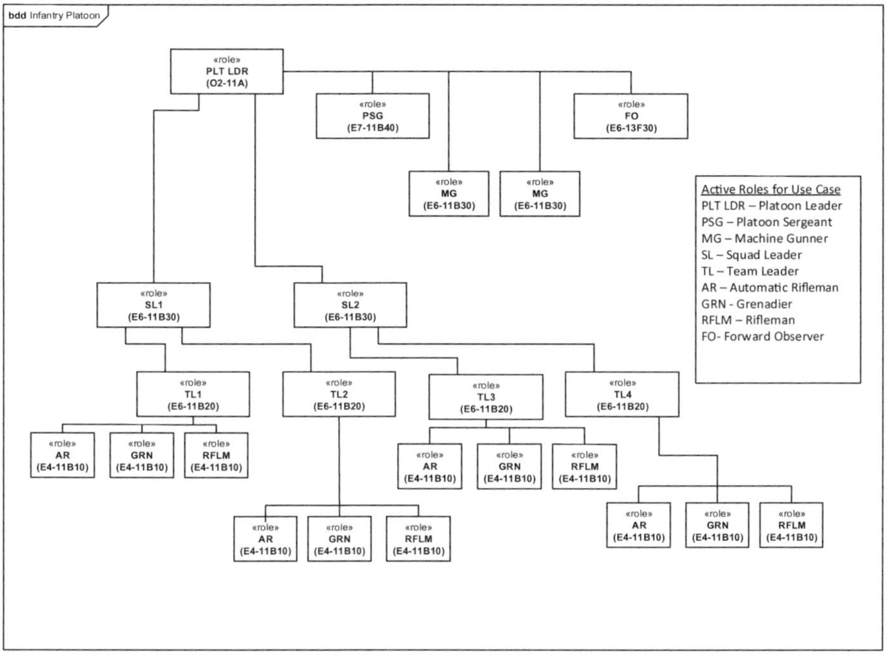

Fig. B.4 Block diagram for Human View roles

B.6 Training

A Block diagram was also used to represent the Training requirements based on the required cognitive and physical competencies. This was rendered as a SysML Internal Block diagram to represent the interface between a role and the assigned tasks. This is shown in Fig. B.5.

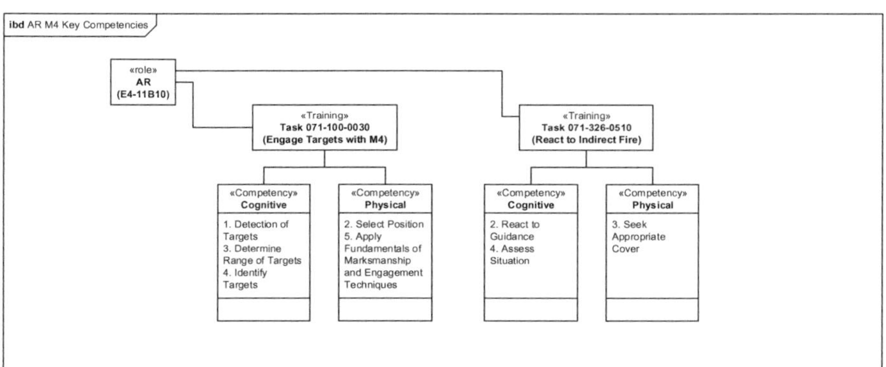

Fig. B.5 Internal block diagram for Human View training

B.7 Human Network

A SysML Sequence diagram can be used to represent the behavior of a process and the information exchanged as roles complete tasks. Figure B.6 represents the Human Networks view using a Sequence diagram. It depicts the sequence of tasks completed by specific roles for "Soldiers under Direct Fire". Time runs from top to bottom of the diagram, and the high-level task process is shown on the left-hand side of the diagram while the individual tasks are referenced by task number under each role. The horizontal lines between tasks are the communications that occur between the roles.

B.8 Metrics

The SysML Requirements diagram is used for the Metrics view, as shown in Fig. B.7. The Requirements diagram represents text-based requirements and their relationship with the other model elements. The requirement is expressed in the top-level block using a text string; the subsequent callouts detail how that requirement can be met through the associated metrics. For the React to Contact scenario, the different rifle types evaluated in the scenario will result in different values for the metrics, impacting the overall mission success requirement.

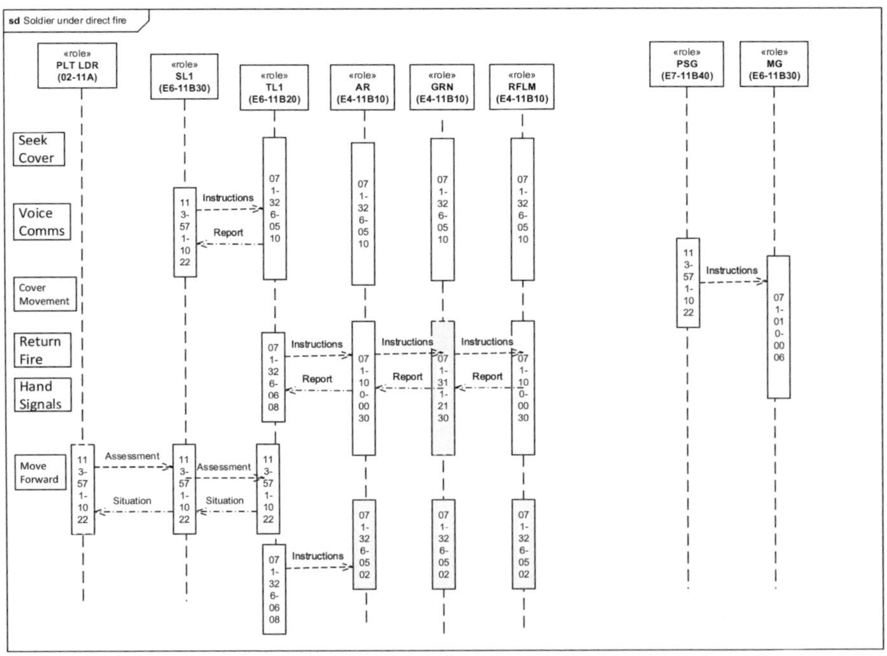

Fig. B.6 Sequence diagram for Human Network view

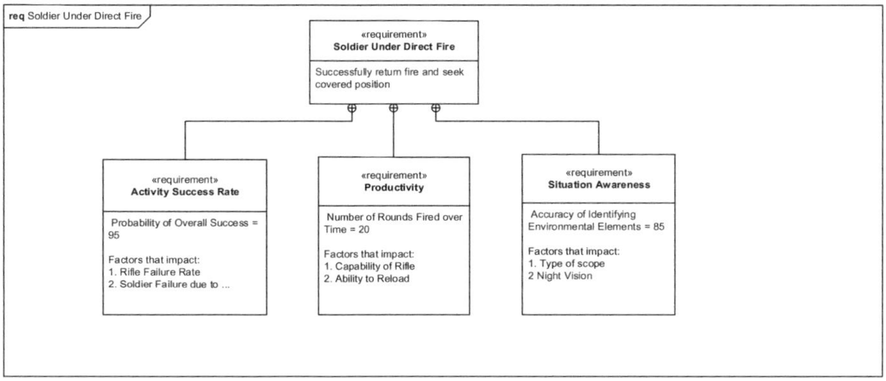

Fig. B.7 Requirements diagram for metrics Human View

B.9 Constraints

The Constraints view is rendered using a SysML Parametric diagram. The Parametric diagram represents real world constraints and is used to support the analysis of stakeholder concerns. Figure B.8 shows the constraints that indicate the limiting physical, cognitive and temporal workload of the soldier.

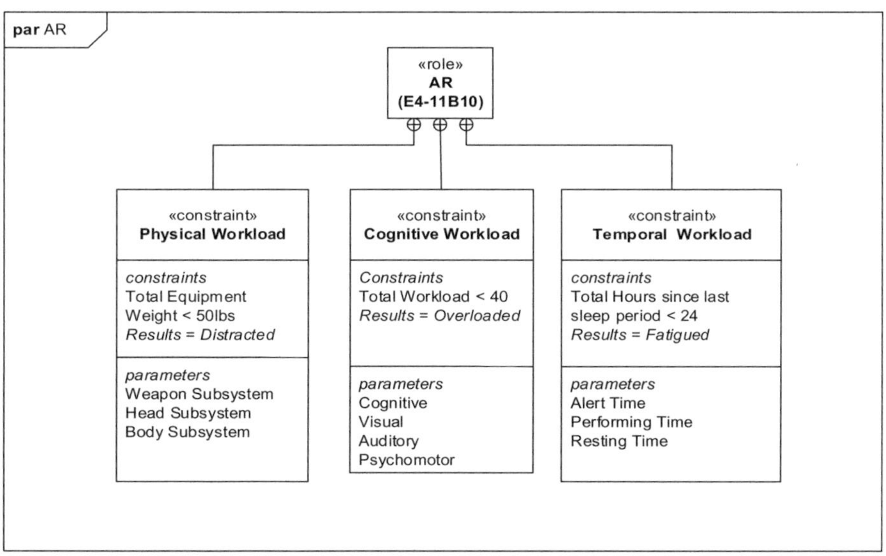

Fig. B.8 Parametric diagram for Human View constraints

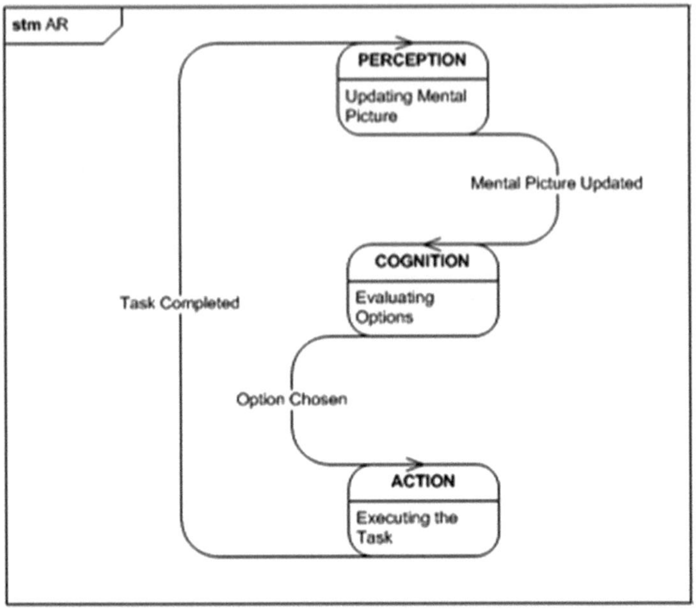

Fig. B.9 State diagram for Human View dynamics

B.10 Dynamics

A representation of soldier behavior can be depicted using a SysML State Machine diagram. The diagram represents behavior in terms of transitions between states triggered by events, as shown in Fig. B.9. In this case the diagram represents a soldier's transition through perception, cognition, and action as a task is executed.

B.11 Conclusion

This example completed a rendering of a full set of Human Views using SysML diagrams. In this case, a different diagram was chosen for each view to exercise the full library of SysML diagrams. However, depending on the view content, different diagrams may be chosen for each model than illustrated here, and the same diagram type may be used for multiple views within the same viewpoint. The SysML implementations shown are at a rudimentary level in order to evaluate the diagram use for the view, and may not fully conform to the SysML standard.

References

Taylor A (2014) React to contact, Use case. Version 4.0, US Army Soldier RDEC, Natick, MA

DOA (2007) The infantry rifleman handbook: common battle drills for all infantry units—TASK: React to contact (07-3-D9501). Department of the Army Field Manual 3-21.8, Washington, DC

MIX
Papier aus verantwortungsvollen Quellen
Paper from responsible sources
FSC® C105338

If you have any concerns about our products,
you can contact us on
ProductSafety@springernature.com

In case Publisher is established outside the EU,
the EU authorized representative is:
Springer Nature Customer Service Center GmbH
Europaplatz 3, 69115 Heidelberg, Germany

Printed by Libri Plureos GmbH
in Hamburg, Germany